UNDERSTANDING OSCILLATORS

2ND EDITION

IRVING M. GOTTLIEB

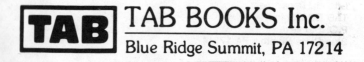

TAB TAB BOOKS Inc.

Blue Ridge Summit, PA 17214

SECOND EDITION

FIRST PRINTING

Copyright © 1987 by TAB BOOKS Inc.

Printed in the United States of America

Library of Congress Cataloging in Publication Data

Gottlieb, Irving M.
    Understanding oscillators.

    Includes index.
    1. Oscillators, Electric. I. Title.
TK7872.07G66 1987   621.3815′33   86-30009
ISBN 0-8306-0715-3
ISBN 0-8306-2715-4 (pbk.)

Questions regarding the content of this book should be addressed to:

    Reader Inquiry Branch
    Editorial Department
    TAB BOOKS Inc.
    P.O. Box 40
    Blue Ridge Summit, PA 17214

Contents

Introduction

1 Frequency-Determining Elements of Oscillators

Parallel-Tuned LC Circuit—Losses in a Tank Circuit—Characteristics of "Ideal" LC Resonant Circuit—Resonance in the Parallel-Tuned LC Circuit—Practical Tank Circuits with Finite Losses—Figure of Merit, "Q"—Physical Interpretation of R_O—Phase Characteristics of Parallel-Tuned LC Circuit—Series-Resonant Tank Circuits—L/C Ratio in Tank Circuits—Transmission Lines—The Delay Line—Distributed Parameters from "Lumped" LC Circuit—Resonance in Transmission Lines—Concept of Field Propagation in Waveguides—Comparison of Lines and Guides—Resonant Cavities—Quartz Crystals—A Closer Look at Crystal Operating Conditions—Magnetostrictive Element—The Magnetostriction Oscillator—The Tuning Fork—RC Networks as Oscillating Elements—LC Networks as Phase Shifters.

2 Active Devices of Oscillators

Switching Actions in Tubes—Amplification—The Class-C Amplifier—Operating Modes of Transistor Amplifiers—Modes of Operation for Op Amps and Logic Circuits—Neon Bulb as a Switching Device—Thyratrons—The Thyratron Inverter—Spark-Gap Oscillation—Negative-Resistance Devices—The Dynatron Oscillator—Transitron Oscillator—The Unijunction Transistor—Triode Input as Negative Resistance—The Saturable Magnetic Core—Oscillation in the Saturable-Core Circuit—The Electron Beam in a Vacuum—The Magnetron—The Reflex Klystron—Traveling-Wave Tubes and the Backward-Wave Oscillator—Oscillator Theory in Terms of the Universal Amplifier—Some Considerations in the Selection of Semiconductor Devices for Oscillators

Introduction

THE SUBJECT OF OSCILLATORS HAS BEEN SOMEWHAT OF A dilemma; on the one hand, we have never lacked for mathematically oriented treatises—the topic appears to be a fertile field for the "long-haired" approach. While these serve the needs of the narrow specialist, they are downright foreboding to the everyday electronics practitioner. On the other hand, one also observes the tendency to trivialize oscillator circuits as nothing more than a quick association of logic devices and resonant circuits. Neither of these approaches readily provides the required insights to devise oscillators with optimized performance features, to service systems highly dependent upon oscillator behavior, or to understand the many trade-offs involved in tailoring practical oscillators to specific demands. Whereas it would be unrealistic to infer that these two approaches do not have their place, it appears obvious that a *third* approach could be useful in bringing theory and hardware together with minimal head-scratching.

This third approach to the topic of oscillators leans heavily on the concept of the universal amplifier. It stems from the fact that most oscillators can be successfully implemented with more than a single type of active device. Although it may not be feasible to directly substitute one active device for another, a little experimentation with the dc supply, bias networks, and feedback circuits does indeed enable a wide variety of oscillators to operate in essentially the same manner with npn or pnp transistors, N-channel of P-channel JFETs, MOSFETs, op amps or IC's, or with electron tubes. Accordingly, this book chooses to deal with basic operating principles predicated upon the use of the universal active-device or amplifier. This makes more sense than concentrating on a specific device, for most oscillator circuits owe no dependency to any single type of amplifying device.

Once grasped, the theory of the general oscillator is easily put to practical use in actual oscillators where concern must be given to the specific active device, to hardware and performance specifications, and to component values. To this end, the final chapter of the book presents numerous solid-state oscillators from which the intelligent hobbyist and practical engineer can obtain useful guidance for many kinds of projects.

It is felt that the reader will encounter little difficulty acclimating to the concept of the universal amplifier, for it is none other than the triangular symbol commonly seen in system block diagrams. Although it hasn't been widely used in conjunction with other circuit symbols, the combination works very well with oscillators. It is respectfully submitted that this book will thereby serve as a unique format for useful information about oscillators.

Chapter 1

Frequency-Determining Elements of Oscillators

A GOOD WAY TO UNDERSTAND OSCILLATORS IS TO VIEW THEM as made up of three essential sections. These are:

1. The frequency-determining section.
2. The active device.
3. A source of dc power.

The validity of this viewpoint does not require that the three sections be physically separate entities. This chapter will treat the characteristics of the elements involved in the frequency-determining section.

PARALLEL-TUNED LC CIRCUIT

Academically and practically, the parallel LC arrangement known as a "tank" circuit is the most important element for us to become familiar with. In its simplest and most frequently encountered form, it is made up of a single inductor and a single capacitor. Whether or not we desire it, the inevitable "uninvited guests," a number of dissipative losses, are always present. (See Fig. 1-1.) In the circuit, these losses behave as resistances. Their presence can, indeed, be closely simulated by simple insertions of resistance into the tank circuit. In Fig. 1-2 we see a possible way in which this can be done. This is the most convenient method and will be

1

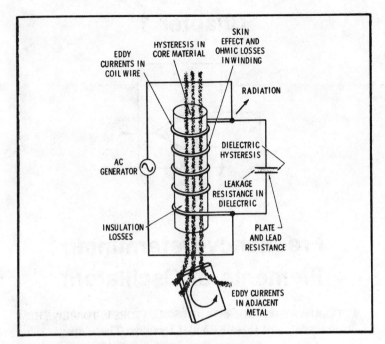

Fig. 1-1. Some possible losses in an LC tank circuit.

used frequently in the equations for computing the various tank circuit quantities.

LOSSES IN A TANK CIRCUIT

Different losses predominate under different situations. In general, the higher the frequency, the greater the radiation loss. Magnetic hysteresis is only of consequence when a ferromagnetic core is used, such as powdered iron. The losses due to eddy cur-

Fig. 1-2. Representation of losses in an LC tank circuit by series resistances R_L and R_C.

2

rents are, in reality, brought about by transformer action in which the offending material constitutes a short-circuited "secondary." This being true, we must expect eddy-current losses in the cross section of the coil winding itself. Skin effect is an alternating-current phenomenon that causes the current to concentrate near the surface of the conductor. This is because the more central regions of the conductor are encircled by more magnetic lines than are the regions closer to the surface (see Fig. 1-3). The more lines of magnetic force encircling a conductor, the greater the inductance of the conductor. Hence, the central regions of a conductor carrying alternating current offer higher inductive reactance to the flow of current.

The higher the frequency, the more pronounced is this effect; that is, the greater the tendency of current to concentrate at or near the surface, thereby reducing the effective cross section of the conductor. Because of skin effect, the resistance offered to the passage of high-frequency current is much higher than the dc resistance. (Inductance does not affect the flow, or distribution of dc.) We are not surprised that skin-effect losses are reduced by using hollow conductors of copper content equal to small gauge wire, but which possess a much greater surface area. Also, stranded wire offers more surface for high-frequency conduction than does its "dc equivalent" in solid wire. Stranded wire with each individual wire insulated (Litz wire) is particularly well suited for the flow of high-frequency current.

(A) Flux around conductor carrying high-frequency current.

(B) Current density at high frequencies.

Fig. 1-3. Magnetic lines of force around a current-carrying wire.

Dielectric hysteresis in insulating materials is the electrostatic counterpart of magnetic hysteresis in magnetic materials. A frictional effect is displayed by the polarized molecules when they are urged to reverse their charge orientation under the influence of an alternating electric field. There are other losses. Those described and those shown in Fig. 1-1 are, however, the most important. Significantly, in many applications, only the losses in the inductor are of practical consequence, for capacitors often have negligible losses from the standpoint of many practical oscillator circuits.

CHARACTERISTICS OF "IDEAL" LC RESONANT CIRCUIT

We find ourselves in a much better position to understand the proprieties of an actual "lossy" tank circuit by first investigating the interesting characteristics displayed by an "ideal" tank circuit in which it is postulated that no losses of any kind exist. It is obvious that such an ideal tank circuit must be made up of an inductor and a capacitor that, likewise, have no losses. In Fig. 1-4, we see depicted the important feature of such ideal elements, i.e., when an ac voltage is impressed across an ideal inductor or an ideal capacitor, current is consumed, but no power is dissipated. Although there is current through these elements, and voltage exists across them, the wattmeters show a zero reading. This may seem strange at first; such a situation is the consequence of the 90° difference in phase between voltage and current. This phase condition is shown in Fig. 1-5 for the ideal conductor, and in Fig. 1-6 for the ideal capacitor.

In both instances, power is drawn from the source for a quarter cycle, but is returned to the source during the ensuing quarter

Fig. 1-4. Voltage and current in an ideal inductor and capacitor.

4

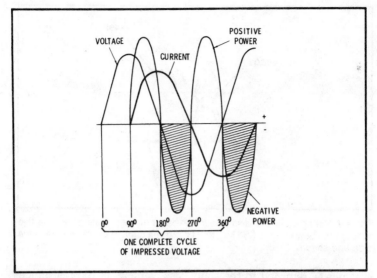

Fig. 1-5. Voltage, current, and power curves for an ideal inductance.

cycle. This makes the power frequency twice that of the voltage or current waves. This need not be cause for surprise, since the same situation prevails for a resistance energized from an ac source. It turns out that the double-frequency power curve is of little practical consequence as such. Of great importance is the fact that the negative portions of the power curves in Figs. 1-5 and 1-6 repre-

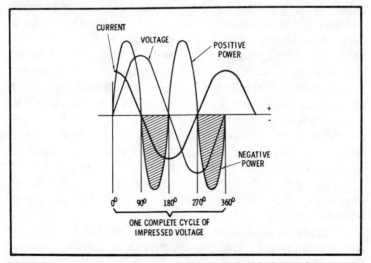

Fig. 1-6. Voltage, current, and power curves for an ideal capacitance.

5

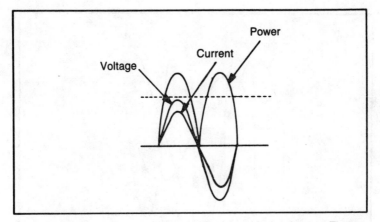

Fig. 1-7. Voltage, current, and power curves for an ideal resistance. The two loops of the power curve (the largest of the three curves) should be the *same* size and shape.

sent power returned to the source; conversely, in the resistance circuit of Fig. 1-7 we note there are no negative portions of the power curve. (All the power drawn by the resistance is dissipated as heat and/or light; *no* power is returned to the source at any time.)

Negative Power

We observe in Figs. 1-5 and 1-6 that sometimes the voltage is positive when the current is negative and vice-versa. By the algebraic law of signs (the product of quantities having unlike signs yields a negative number) it is just such occurrences that produce the negative excursions of the power waveform. Also, every time either voltage or current crosses the zero axis, the power wave must also cross the zero axis. (Zero times any number is zero.) Inasmuch as the power curve results from multiplying instantaneous voltage by instantaneous current valves, we see why the power curve is twice the frequency of the voltage or current waves. Although negative voltage is every bit as good as positive voltage and despite the fact that the same is true for negative and positive current, this reasoning cannot be extended to explain the physical significance of negative power. Positive power is the power taken from the ac source by the load; negative power is power returned to the source from the load.

Although the ideal inductor and the ideal capacitor do not, themselves, consume or dissipate power, the current that they cause to flow in the line can and does cause power loss in the resistance

6

of the line or connecting conductors and in the internal resistance of the source. We see that it still holds true that current flowing through a resistance causes power loss. Thus, although our ideal elements would produce no power loss *within themselves,* their insertion in a circuit must, nevertheless, cause power loss within other portions of the circuit. These matters are fundamental and should be the subject of considerable contemplation before going on. (We reflect that a flywheel, a rotating mass behaving in an analogous way to inductance, consumes no power from the engine. Also, an ideal, that is, frictionless, spring returns all of the mechanical power used to deflect it.)

Performance of Ideal Tank Circuit

We are now better prepared to consider the performance of the ideal tank circuit formed by connecting the ideal inductor and ideal capacitor in parallel. Such a tank circuit is shown in Fig. 1-8. Let us suppose that the generator delivers a frequency equal to the resonant frequency of the LC combination. Resonance in such a circuit corresponds to that frequency at which the reactance of the inductor and the reactance of the capacitor are equal, but of opposite sign. From the individual properties of ideal inductors and ideal capacitors, we should anticipate that no power would be consumed from the source. This is indeed true. However, we would also find that, *at resonance,* no current is drawn from the line either. Can we correctly infer from such a situation that the source is *not* actually needed to sustain oscillations in the ideal tank circuit?

From a theoretical viewpoint such a conclusion is entirely valid. The switch in Fig. 1-8 could be opened and large currents would circulate in oscillatory fashion between the ideal inductor and the ideal capacitor. Our ideal tank circuit would now be self-oscillatory

Fig. 1-8. A resonant tank composed of ideal elements.

at its resonant frequency. We would have a sort of perpetual motion, but still not of the variety attempted by inventors unversed in physical law. That is, the ideal tank circuit, though self-oscillatory, could not long supply power to a load; as soon as we *extracted* power from such a tank circuit, we would effectively introduce resistance, thereby destroying its ideal nature. An "ideal" pendulum involving no frictional losses whatsoever would swing back and forth through eternity; however if we attempted to harness the motion of the rod to perform mechanical work of some sort, we would dissipate its stored energy, thereby damping the amplitude of successive swings until oscillation ceased entirely.

RESONANCE IN THE PARALLEL-TUNED LC CIRCUIT

We saw in Fig. 1-5 that the current in an ideal inductor lags the applied voltage by a quarter cycle, or $90°$. We saw in Fig. 1-6 that the current in an ideal capacitor leads the applied voltage by a quarter cycle, or $90°$. Significantly, at some frequency, the current in the line feeding a parallel combination of ideal inductor and ideal capacitor will neither lead nor lag the applied voltage, but will be in phase with it. The frequency must be such that inductive reactance and capacitive reactance are equal numerically, for only then can exact cancellation of phase displacement between current and applied voltage occur. At other frequencies, the phase displacement between current and applied voltage will be less than $90°$ lead or lag, but cannot be zero.

We may think of inductance and capacitance as tending to cancel each other's power to cause phase shift between current and applied voltage. Inductive reactance, X_L increases as frequency is increased ($X_L = 2\pi fL$). Capacitive reactance X_C decreases as frequency is increased: [$X_C = 1/(2\pi fC)$]. Therefore, at the one frequency known as the *resonant frequency*, and at this frequency only, the two reactances are numerically equal. See Fig. 1-9. Inasmuch as the phase leading and lagging effects of the two reactances then cancel, the tank circuit no longer behaves as a reactance, but rather as a pure resistance. In the ideal tank circuit, this resistance would be *infinite* in value; in practical tank circuits, we shall find that this resistance, R_0, has a finite value dependent upon the inductance, capacitance, and resistance in the tank circuit.

Inductance/Capacitance Relationships for Resonance

At resonance X_L and X_C are equal. Therefore we may equate

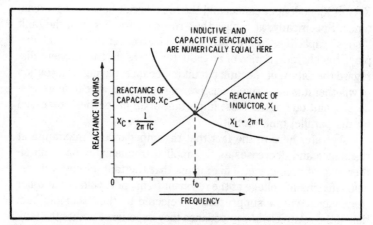

Fig. 1-9. Variation of inductive and capacitive reactance with frequency.

their equivalents thus, $2\pi f_0 L = 1/(2\pi f_0 C)$. If we next make the algebraic transpositions necessary to solve this identity for the resonant frequency, f_0, we obtain $f_0 = 1/(2\pi\sqrt{LC})$. We see from this equation that for a given resonant frequency, f_0, *many* different combinations of L and C are possible; we may resonate with large inductance and small capacitance, or with the converse arrangement (see Fig. 1-10). Above and below the frequency of resonance,

Fig. 1-10. Different combinations of L and C for obtaining the same resonant frequency.

X_L and X_C no longer cancel and the tank circuit behaves as a reactance. Specifically, at higher than resonant frequencies, the tank "looks" capacitive; at lower than resonant frequencies the tank displays the characteristics of an inductor. If we for the moment disregard the nature of the tank circuit impedance, paying no attention to whether it is inductive, resistive, or capacitive, we can more easily pinpoint one of the most important of all properties possessed by the parallel tank circuit.

We refer here to the fact that the impedance is maximum at resonance and decreases as the applied frequency departs on either side of resonance. This means that the tank circuit will develop maximum voltage at the resonant frequency; voltages at other frequencies will be suppressed or rejected by the "shorting" action of the relatively low impedance they experience across the tank circuit terminals. In the ideal tank circuit, the resonant frequency is supported freely across a resistive impedance infinitely high; frequencies only slightly higher or lower than the resonant frequency experience a short circuit and are thereby rejected. Practical tank circuits gradually discriminate against off-resonant frequencies (see Fig. 1-11). The better they do this, that is, the greater the rejection for a given percentage of departure from true resonance, the greater we say is the selectivity of the tank circuit. In Fig. 1-11, the resistance R is not part of the tank circuit, but is used to isolate the tank circuit from the source. Otherwise, a low impedance source could mask the impedance variations of the tank circuit.

The following conditions at resonance of a parallel LC tank cir-

Fig. 1-11. Typical frequency response of a parallel LC circuit.

Fig. 1-12. Resonance conditions in a typical LC tank circuit.

cuit (R_L and R_C are small) apply in Fig. 1-12:

    1. The applied voltage, V, and the line current, I_{LINE}, are in phase.

    2. Inductor current, I_L, and capacitor current, I_C, are of equal value because inductive reactance and capacitive reactance are equal.

    3. The voltage, V, applied across the tank circuit rises to its maximum value because the impedance of the tank circuit attains its highest value.

    4. The tank circuit behaves as a pure resistance.

    5. The line current, I_{LINE}, attains its minimum value.

PRACTICAL TANK CIRCUITS WITH FINITE LOSSES

    We have been dealing with *ideal* reactive elements and *ideal* tank circuits in order to increase our insight into the nature of actual tank circuits. We realize that practical inductors and capacitors contain losses. Let us now investigate the effect of these losses in the actual tank circuit. First, we can no longer expect zero line current at resonance. Although the line current will be minimum at resonance, a residual value will exist. This must imply that the terminal impedance at resonance is no longer of infinite value (see Fig. 1-12). In order to acquire a working acquaintance with tank

11

circuits, it is expedient that we learn how the several relevant properties of the tank circuit are interrelated. To make such a study meaningful, we must understand the significance of the figure of merit known as the "Q" of a tank circuit.

There are several different ways to think of Q. However, notwithstanding the definition best suited to a particular tank circuit application or calculation, Q is basically a measure of the ratio of energy stored per cycle to the energy dissipated per cycle. We see that the Q of the ideal tank circuit, which contains *no* power dissipating resistance or other losses, is infinite. As a corollary of this basic concept of Q, we can also say that Q provides a measure of the frequency selectivity, that is, the bandwidth of the tank circuit (see Fig. 1-13). Q enables us to judge the ability of a tank circuit to suppress harmonics. From the mathematical viewpoint, Q constitutes an extremely convenient means of tying together the various electrical parameters of the tank circuit. In this book, we shall assume that Q always has a value of ten or more. When tank circuit losses are great enough to reduce Q below this value, the simple formulas for resonance and for other properties of the tank circuit are no longer reasonably accurate and the resonance curve

Fig. 1-13. Resonance curves showing effect of Q on frequency selectivity.

12

Fig. 1-14. Parallel LC tank circuit for use with Charts 1-1, 1-2, and 1-3.

loses its symmetry. In Fig. 1-13, when the sum of R_C and R_L is low, the Q is high. When the sum of R_C and R_L is high, the Q is low. It fortunately happens that the vast majority of LC networks intended to function as tank circuits have Q's exceeding ten. Q's of twenty to one hundred are commonly found in receivers. Q's of several hundred may be found in transmitters. Quartz crystals and microwave cavities commonly have Q's of several tens of thousands and the Q of special quartz crystals may be on the order of a half million.

FIGURE OF MERIT, "Q"

We will concern ourselves primarily with the value of Q that exists at (or very close to) resonance. It will be convenient to designate Q for this condition as "Q_0." A great deal of insight into the parallel LC tank circuit can be gained by a serious study of the "Q_0" relationships shown in Chart 1-1 (also see Fig. 1-14). The fact that Q_0 can be expressed in so many ways is, in itself, indicative of the importance of this tank-circuit parameter. We see that a knowledge of Q and one quantity other than f_0 is sufficient to enable us to start a chain of calculations from which we can determine *all* of the many parameters associated with the parallel LC tank circuit.

PHYSICAL INTERPRETATION OF R_0

R_0 is the impedance seen across the tank terminals at resonance. Significantly, R_0 appears as a pure resistance. It is interesting to contemplate that, although R_0 is purely resistive, its value

13

Chart 1-1. Relationships of the
Resonant Q Factor (Q_0) in the Parallel LC Tank

1. $Q_0 = \dfrac{R_0}{(2\pi f_0)(L)}$

   Where R_0 is the impedance across the tank terminals at resonant frequency f_0.

2. $Q_0 = (R_0)(2\pi f_0)(C)$

3. $Q_0 = \dfrac{1}{(2\pi f_0)(C)(R_S)}$

   Where R_S is the sum of the losses in the inductor and capacitor expressed as equivalent series resistances. That is, $R_S = R_L + R_C$ in Fig. 1-12. In many practical cases, we may reasonably say that $R_S = R_L$, R_C being negligible.

4. Thus, $Q_0 = \dfrac{1}{(2\pi f_0)(C)(R_L)}$

   When R_C is negligible.

5. $Q_0 = \dfrac{(2\pi f_0)(L)}{R_L}$

   It is also true that $Q = \dfrac{(2\pi f)(L)}{R_L}$, that is, the Q of the inductor at any frequency, f is obtained by substituting f for f_0. (Inasmuch as losses vary with frequency, R_L is likely to be a different value if f differs appreciably from f_0.)

6. $Q_0 = \dfrac{(0.5)(f_0)}{f_0 - f_x}$

   Where f_x is the frequency below or above resonance which causes the impedance (or voltage) across the tank terminals to be 71% of the impedance (or voltage) at resonance.

7. $Q_0 = \dfrac{(2\pi)(\text{energy stored per cycle})}{(\text{energy dissipated per cycle})}$ or, for practical purposes,

   $Q_0 = \dfrac{\text{volt-amperes}}{\text{watts}}$ which is obtained from $\dfrac{(V)(I_{LINE})}{W}$

8. Also, $Q_0 = \dfrac{I_L}{I_{LINE}} = \dfrac{I_C}{I_{LINE}}$ when the tank is at resonance.

9. In Q_0/π cycles the oscillating voltage across a shock-excited tank circuit falls to 37% of its maximum value if additional disturbances are not applied to the tank circuit.

10. In $Q_0/2\pi$ cycles, the stored energy in a tank circuit falls to 37% of its maximum value unless replenished before elapse of that time. We imply here, the time interval following a pulse or disturbance, or after disconnection of the ac source.

11. Significance of $Q_0 = \frac{1}{2}$ or less

    When sufficient resistance is introduced to make $Q_0 = \frac{1}{2}$, a resonant circuit is critically damped; no oscillation can be provoked by transients or shock excitation. This condition prevails when the tank circuit is overdamped by making Q_0 less than $\frac{1}{2}$.

is largely governed by *reactances*. This is revealed by the two relationships:

$$R_0 = Q_0 \; (2\pi f_0) \; (L)$$

and

$$R_0 = \frac{Q_0}{(2\pi f_0) \ (C)}$$

In these relationships, we note that $(2\pi f_0)$ (L) is inductive reactance and $1/ (2\pi f_0)$ (C) is capacitive reactance, both reactances being designated for the resonant frequency f_0.

Chart 1-2 is included to depict the numerous parallel LC tank calculations that involve the resonant value of Q, that is, Q_0. Chart 1-3 is a list of other useful formulas relating various parameters of the parallel LC tank circuit. In these three tables, resistances such as R_0, R_L, R_C, and R_S must be expressed in ohms. Reactances, such as X_L and X_C, must also be expressed in ohms. Frequencies such as f, f_0, or f_x must be expressed in hertz. The same is true of bandwidth, BW. Inductance, L, must be expressed in henrys, and capacitance in farads. Finally, Q and Q_0 are dimensionless numbers designating the figure of merit of the tank circuit.

PHASE CHARACTERISTICS OF PARALLEL-TUNED LC CIRCUIT

Before we leave the parallel LC tank circuit, we should visualize its effect upon the phase relationship between impressed voltage and resulting line current. Figure 1-15 illustrates the general nature of phase response in such a circuit. Significantly, we see that at resonance, there is no phase shift. Below the resonant frequency, the circuit acts as an inductor in parallel with a resistance. Consequently the tank "looks" inductive. The further we depart from resonance in the frequency region below resonance, the

Chart 1-2. Calculation of Parallel LC Tank Circuit Parameters with the Aid of the Resonant Q Factor, Q_0

1. $R_0 = (Q_0)(2\pi f_0)(L)$

2. $R_0 = \dfrac{Q_0}{(2\pi f_0)(C)}$

3. $L = \dfrac{R_0}{(Q_0)(2\pi f_0)}$

4. $C = \dfrac{1}{(Q_0)(R_S)(2\pi f_0)}$

5. $C = \dfrac{1}{(Q_0)(R_L)(2\pi f_0)}$ *

6. $R_S = \dfrac{(2\pi f_0)(L)}{Q_0}$

7. $L = \dfrac{Q_0(R_S)}{2\pi f_0}$

8. $R_L = \dfrac{(2\pi f_0)(L)*}{Q_0}$

9. $L = \dfrac{(Q_0)(R_L)*}{2\pi f_0}$

10. $I_L = I_C = (Q_0)(I_{LINE})$

*When capacitor losses are negligible.

1. $X_L = X_C = \sqrt{\dfrac{L}{C}}$ This relationship holds true only at resonance.

2. $BW = \dfrac{1}{(2\pi)(R_0)(C)}$ These formulas enabled us to calculate the bandwidth of a parallel LC tank circuit. B W is the bandwidth in hertz where the impedance, or voltage, is 71% of the value at resonance.

3. $BW = \dfrac{R_S}{(2\pi)(L)}$

4. $X_L = (2\pi f)(L)$ Where X_L is the reactance of the inductor at any frequency, f.

5. $X_C = \dfrac{1}{(2\pi f)(C)}$ Where X_C is the reactance of the capacitor at any frequency, f.

6. $f_0 = \dfrac{1}{2\pi\sqrt{LC}}$

7. $R_0 = \dfrac{L}{(R_S)(C)}$

greater the effect of the inductance and the less the effect of the resistance. (The effective parallel resistance is of the same nature as R_0 but at a different frequency.) At low frequencies, the phase shift approaches ninety degrees and the tank appears therefore as a more nearly perfect inductor. Above the resonant frequency, the circuit acts as a capacitor in parallel with a resistance. (The resis-

Fig. 1-15. Typical phase response of a parallel LC tank circuit.

tance is again of the same nature as R_0, but corresponds to a frequency higher than f_0.) At high frequencies, the phase shift approaches ninety degrees, but in the opposite direction as in the low frequency case. The tank appears therefore as a more nearly perfect capacitor.

SERIES-RESONANT TANK CIRCUITS

In oscillators, and in the equivalent circuits of oscillating elements, we often encounter the *series-resonant* combination of inductance and capacitance, and, of course, the inevitable resistance. Such a configuration also behaves as a "tank," that is, its properties are the manifestations of energy storage. The characteristics displayed by the series-tuned tank circuit are, generally speaking, opposite to those of the parallel-tuned tank circuit. In the parallel tank circuit, we found that the line current attained its minimum value at resonance. Conversely, in the series tank circuit, line current is maximum at resonance. In a series tank circuit composed of a lossless inductor and a lossless capacitor, the line current would reveal the presence of a perfect short circuit placed across the ac source. In practical series tank circuits, line current at resonance is limited only by the equivalent series resistances of the inductor and capacitor, which are, respectively, R_L and R_C.

Q in Series Tank Circuits

Our concepts of Q and Q_0 are, in principle still valid for the series tank circuit; however we must now use a different set of formulas to relate the Q valve to the various parameters of the series tank circuit. The important relationships in the series tank circuit are given in Chart 1-4. As previously was the case, we must stipulate a minimum Q_0 of ten in order to preserve the validity of these formulas. In most applications encountered in practice, this will not restrict our ability to visualize the operation of the series tank or prevent us from calculating or estimating reasonably accurate parameter values, for a Q_0 below ten is the exception rather than the rule.

Resonance in Series-Tuned LC Circuit

We observe that the resonance curve is essentially of the same nature as in the parallel tank; however we now deal with *line current* rather than applied voltage (or impedance) as a function of fre-

Chart 1-4. Parameters of the Series Tank Circuit

1. $Q_0 = \dfrac{(2\pi f_0)(L)}{R_S}$

2. $Q_0 = \dfrac{(2\pi f_0)(L)}{R_L}$ Where capacitor losses R_C are much less than inductor losses R_L.

3. At resonance $V_L = V_C = Q_0 V_{LINE}$

4. $R_r = \dfrac{V_0}{I_0}$ Where R_r is the effective resistance of the circuit seen by the ac source when resonance exists. R_r in the series tank corresponds to R_0 in the parallel tank.

    Where V_0 is the terminal voltage at resonance.

    Where I_0 is the line current at resonance.

5. $R_r = R_S = R_L + R_C$

6. $Q_0 = \dfrac{L}{(2\pi f_0)(C)(R_S)}$ When R_C is negligible.

    $= \dfrac{L}{(2\pi f_0)(C)(R_L)}$

7. $I_0 = \dfrac{V_0}{R_S}$

quency (see Fig. 1-16). At the resonance frequency, f_0, the line current is restricted by the equivalent series resistance, R_S, representing the losses in the inductor, R_L, and the losses in the capacitor R_C. When we inspect the phase response of the series tank in Fig. 1-17, we see that it is essentially the converse of the phase response associated with the parallel tank circuit shown in Fig. 1-15. Whereas the parallel tank circuit is inductive at frequencies below resonance, we find the series circuit is capacitive in this frequency region. Similarly, at frequencies above resonance, the series circuit is inductive rather than capacitive as we found to be the case in the parallel tank circuit. It is interesting to note that the voltages existing across the inductor and capacitor of the series circuit can be many times the value of the line voltage impressed across the circuit terminals (see Fig. 1-18). We recall the counterpart of this phenomenon in the parallel tank circuit wherein the circulating current, i.e., I_L and I_C, can be many times the value of the line current, I_{LINE}.

L/C RATIO IN TANK CIRCUITS

Of considerable practical importance is the effect of the inductance-to-capacitance ratio on Q_0 in tank circuits. In both parallel and series tank circuits, a given resonant frequency, f_0, can be attained by numerous combinations of inductance and capacity

18

Fig. 1-16. Typical frequency response of a series LC tank circuit.

(see Fig. 1-10). In the parallel tank circuit, for a given value of R_0 and f_0, we increase Q_0 as we make the capacitance larger and the inductance smaller. However, in the series tank circuit, Q_0 is increased, for a given value of R_S and f_0, by making the capacitance

Fig. 1-17. Typical phase response of a series-tuned LC circuit.

Fig. 1-18. Voltage step-up in a resonant series tank circuit.

smaller and the inductance larger. Both the parallel tank circuit and the series tank circuit are prevented from having an infinite Q_0 by the presence of R_S. Whereas high capacitance is necessary to make Q_0 high in the parallel tank circuit, high inductance is necessary to make Q_0 high in the series tank circuit. These statements are mathematically true, but in practice we cannot increase either the capacitance in the parallel tank, or the inductance in the series tank indefinitely without incurring losses, which ultimately bring us to a point of diminishing returns, wherein Q_0 begins to decrease rather than increase.

For example, any resonant frequency, f_0, can be attained in a parallel tank by means of a one-turn coil and an appropriate capacitor. However, as we make f_0 lower in frequency, the capacitor required to achieve resonance becomes larger. The dielectric losses in a physically large capacitor would mount up until further expansion of its physical size would add sufficient losses to cause subsequent decrease of Q_0 with any further pursuit of this trend.

A similar argument can be presented for the inductor in the series tank. Thus, in practical tank circuits of both varieties, we find that we are restricted to a Q_0 value, which cannot be exceeded by further change in the ratio of inductance to capacitance. Furthermore, in the parallel tank circuit, the change in the value of R_0 with the LC ratio introduces other difficulties. Often, the parallel tank is required to have an R_0 value within a certain range. This may conflict with the Q_0 requirement and a compromise must be made. Also, we should appreciate the fact that an extremely high Q_0 is not always desirable. The selectivity of a high Q_0 tank circuit can be sufficient to discriminate against the frequency spectrum required in voice modulation.

In other instances, the sharp tuning of the high Q tank may result in critical adjustments that are not mechanically stable. In transmitters, diathermy, and induction and dielectric heating equipment, too high a Q produces excessive circulating current in parallel tank circuits, thereby dissipating power and generating heat where least desired. Finally, we note that in both types of tank circuits, it is possible to have losses and a high Q_0 simultaneously. For example, series-mode resonance in a quartz crystal is accompanied by a relatively high value of R_s. However, Q_0 is nevertheless very high due to the high *ratio* of effective L to R_s.

TRANSMISSION LINES

Transmission lines, such as parallel wires or coaxial cable, are capable of producing resonance phenomena similar to that associated with "lumped" inductance and capacitance elements. The inductance and capacitance are present in the transmission line, as in the coil and capacitor, but are distributed along the length of the line. If the end of the line farthest from the generator is either open-circuited or short-circuited, resonances will exist at distances corresponding to quarter-wavelength spacings from that end. From the viewpoint of the generator, that is, in terms of the line current it is called upon to deliver, the line alternates from series to parallel resonance each time we change the line length one quarter wavelength. This is shown in Fig. 1-19 in another manner.

Fig. 1-19. Parallel- and series-type resonances in a short-circuited line.

We see that series-type and parallel-type resonances occur at successive quarter-wave distances from the shorted end of the line. An open line of any number of odd wavelengths long behaves similarly, that is, it undergoes a similar type of resonance, to a shorted line of any number of even wavelengths long. For our study of oscillators, we shall be primarily interested in parallel resonance of transmission line. In Fig. 1-20 we see the simplest line configurations for simulating parallel resonance. At the terminals of the two configurations, voltage is maximum and current is minimum, as at the terminals of a parallel LC tank circuit.

In speaking of a resonant frequency of lumped tank circuits, we found ourselves limited to exactly *one* such frequency for any given LC combination. However, a transmission line can exhibit resonances at many frequencies corresponding to various multiplies of a quarter wavelength. For example, a line that is shorted at its far end, and that is a quarter wavelength at f_1 displays series and parallel-type resonances respectively at $2f_1$, $3f_1$, $4f_1$, $5f_1$, etc. The series resonances do not cause oscillation in oscillators designed to use the properties of parallel resonance. The parallel resonances can result in oscillation at other than the desired frequency, but in practice, it is generally not difficult to ensure oscillation at only the intended frequency. One of the factors that tends to discourage oscillation at such frequencies as $3f_1$, $5f_1$, $7f_1$, etc., in the shorted

Fig. 1-20. Simplest resonant lines for simulating the properties of the parallel-tuned tank circuit.

line (or at $2f_1$, $4f_1$, $6f_1$, etc., in the open line) is that the losses become higher at the multiple frequencies of f_1. Other things being equal, an oscillator will "prefer" to operate at the frequency involving lowest losses, i.e., highest tank-circuit Q.

THE DELAY LINE

The delay line is an important element for certain oscillators of the relaxation type. In these applications, *pulse duration*, rather than pulse repetition rate or frequency, is the controlled parameter. Delay-line stabilized oscillators are used quite extensively in radar and in digital techniques. The delay line is essentially a transmission line. Nevertheless, its operating mode, and generally its physical appearance also, differ from the lines employed as high-Q resonating elements in sinusoidal oscillators.

The simplest delay line comprises a length of open wire or coaxial transmission line. Pulses introduced at one end require a precise time interval to reach the opposite end. If the far end of the line is terminated in a resistance equal to the characteristic impedance of the line, the pulse dissipates its energy in the resistance. If the far end of the line is either open circuited or short circuited, the pulse undergoes reflection and is thereby returned to the source end of the line. Under such conditions, a more efficient use is made of the lines because the pulse is forced to traverse its length *twice*. Thus, the obtainable time delay is twice what it would be if the pulse energy was absorbed in a resistor at the far end of the line (see Fig. 1-21).

If a line is open circuited at its far end, the returned voltage pulse will have the *same* polarity as the initiating pulse. If the line is short circuited at its far end, the returned voltage pulse will have *opposite* polarity with respect to the initiating pulse voltage. The latter case is particularly useful, for the arrival of the reflected pulse is readily utilized to *terminate* the conduction state of the oscillator responsible for the generation of the leading edge of the pulse. In this way a precise pulse duration is established, being determined by the *electrical length* of *the line* rather than by time constants of resistances and capacitances associated with the oscillator. Variations in power supply voltage and tube or transistor temperature cannot exert appreciable effect upon pulse duration when an oscillator is stabilized by a delay line.

An electrical disturbance propagates along typical transmission lines at speeds between 75% and 98% the speed of light in free space, depending upon the dielectric employed for line spacing. A

Fig. 1-21. Delay lines.

delay of one microsecond would require a line length on the order of a city block! Thus, except when delays of a small fraction of a microsecond suffice, the ordinary transmission line would assume impractical physical proportions. Fortunately, it is feasible to construct compact *networks* which simulate transmission line operation, but at relatively slow speeds of pulse propagation (long delay times).

The Artificial Transmission Line

From electric wave filter theory, it is known that the low-pass filter configuration behaves in many respects as a transmission line. A few inductances and capacitances can be connected in such a network to provide a physically compact artificial transmission line from which relatively long delay times can be attained. Three important operating parameters are associated with the artificial transmission line. These are: t, the delay time per section (such a network generally consists of a number of cascaded elementary filter sections), the cutoff frequency, f_0, and the characteristic impedance, Z_0. All three of these parameters are governed, although in different ways, by the values of inductances and capacitances. Other things being equal, pulse fidelity tends to be better as the cutoff frequency is made *higher*, and also as more elementary, or "prototype" sections are cascaded. For proper operation, the characteristic impedance should be nearly the same as the internal generator resistance of the oscillator during pulse production. These requirements impose contradictory design approaches and the construction of delay lines has become a competitive art for specialists. One of the criterions of performance of these devices is the *ratio of pulse delay time to pulse rise time.* The "goodness" of a delay line increases as this ratio becomes higher.

Ⓐ *Transistor blocking oscillator.*

Ⓑ *Tunnel-diode relaxation oscillator.*

Fig. 1-22. Stabilization of wave duration by use of delay lines.

Delay-Line Stabilized Blocking Oscillator

The repetition rate of the blocking oscillator shown in Fig. 1-22A is primarily governed by the resistance-capacitance combination, R1C1. The pulse duration, however, is precisely determined by the delay line. The electrical length of the delay line corresponds to a shorter period of time than the pulse would endure in the absence of the delay line. When the blocking oscillator commences its "on" state, the emitter of the pnp transistor becomes positive with respect to its base (and ground). This initiates propagation of a positive-going wavefront down the line. Reflection occurs at the shorted (grounded) end of the line and an inverted voltage pulse is propagated back toward the emitter end of the line. When the wavefront of this reflected pulse arrives at the emitter, the transistor is deprived of forward conduction bias. This causes the blocking oscillator to abruptly switch to its "off" state, thereby terminating the duration of its generated pulse.

Delay-Line Stabilized Tunnel-Diode Oscillator

In Fig. 1-22B a simple but very useful oscillator is shown. Relaxation oscillations occur, but with the pulse duration controlled by the shorted delay line. It will be noted that the delay line is used in place of the inductor ordinarily associated with this circuit. The value of R must be high enough to permit switching transitions to occur between characteristic curve points shown in the accompanying illustrations.

DISTRIBUTED PARAMETERS
FROM "LUMPED" LC CIRCUIT

In order to bridge mentally the transition between "lumped" tank circuits and resonant lines, a somewhat different conceptual approach is needed. This may be acquired with the aid of Fig. 1-23, which shows how a resonant line can be theoretically evolved from lumped elements. At Fig. 1-23A several relatively large inductors and capacitors are connected to simulate roughly the condition in a resonant line. An ac generator connected to the left pair of terminals will deliver a current which will undergo cyclic maxima and minima as the frequency is varied. Maximum current, from the viewpoint of the generator, corresponds to series resonance; conversely, minimum current corresponds to the resonance associated with the simple parallel-tuned LC tank. For either type of resonance, voltages and currents will now be distributed throughout the net-

A A few large inductors and capacitors

B Moderate number of smaller inductors and capacitors

C Many small inductors and capacitors

Capacitance between the two inductors now substitutes for physical capacitors

D Continuous inductance and capacitance

Self-inductance of wires now substitutes for physical inductors

E Continuous inductance and capacitance due entirely to self-inductance of straight wires and capacitance between them.

Fig. 1-23. Synthesis of a resonant transmission line from lumped elements.

work, rather than concentrated at single circuitry junctions as in a simple tank circuit.

For example, the voltage across alternate capacitors will be very nearly the same. We have simulated the alternate series and parallel type resonance displayed at quarter-wave intervals along a line. At Fig. 1-23B we reduce the size of the inductors and capacitors, but employ more of them. At Fig. 1-23C this process is carried still further. At Fig. 1-23D we arrive at a continuous inductance and no longer rely for inter-line capacitance on physical capacitors. The required capacitors are now so small and are spaced so close together that the inherent capacitance existing *between* the "lines" serves the required function. Finally, at Fig. 1-23E we stretch out the helical coil and rely upon the self-inductance of the straight wires to contribute the necessary inductance.

RESONANCE IN TRANSMISSION LINES

We will be primarily interested in the resonant conditions of lines. These are most effectively displayed by lines that are either short-circuited or open-circuited at their far ("receiving") ends. However, additional insight into the nature of such lines is attainable by comparing them further to lumped tank circuits. One such comparison involves termination by a resistance value that prevents resonances. The value of such a terminating resistance is given by the formula:

$$R = \sqrt{\frac{L}{C}}$$

where,
   R is the resistance in ohms,
   L is the inductance in henrys per unit length of line,
   C is the capacitance in farads per unit length of line.

Interestingly, a similar situation prevails for lumped tanks. In Fig. 1-24 we see that the destruction of resonances in both instances is brought about by resistance values equal to $-\sqrt{L/C}$. In the line, this value is called the characteristic impedance of the line. The characteristic impedance of the line is determined by wire diameter, wire spacing, and the dielectric constant of the insulating material. Coaxial lines operate in the same way as open lines but have the advantage that the external surface of the outer conductor may be operated at ground potential. This produces effective shielding that prevents radiation. The Q is thereby increased, and interferences and feedback troubles are virtually eliminated. The highest Q in coaxial lines is attainable when the ratio of conductor diameters is 3.6 to 1. This results in a characteristic impedance of 77 ohms.

CONCEPT OF FIELD PROPAGATION IN WAVEGUIDES

The evolution of "lumped" circuitry into spatially distributed reactance, as illustrated in Fig. 1-23B aids our visualization of the transmission line. However, the transmission line is not the ultimate configuration; we can carry the process a step further. This

Fig. 1-24. Termination of line by resistance equal to its characteristic impedance destroys resonance completely.

is best approached by considering the coaxial transmission line. Suppose we make the central conductor smaller and smaller in diameter. What must ultimately happen? We can expect an increase in characteristic impedance and perhaps a change in the losses. However, we have nothing to correlate with our experiences with "wired" circuits to help us ascertain the result of eliminating the central conductor altogether. We might intuitively conclude that this would end the usefulness of the cable as a means of conveying high-frequency energy. Such a deduction would indict us as victims of the commonly believed concept that physical conductors are necessary to provide a means of "go" and "return" for high-frequency currents. Such is not actually the case.

The physicist and mathematician view the phenomenon in a somewhat differing light. They do not refute the notion that a varying magnetic and electric field accompany the flow of current in conductors; however, they do subscribe to the notion that the flow of alternating current in such a "circuit" as a transmission line, is the result of an alternating magnetic and electric field propagated down the line, that is, between the two conductors. If this is true, why not eliminate the need for supplying the energy by ohmic contact? Rather, let us deliver energy to the line by means of a small antenna or loop so that the energy is initially supplied in the form of fields. If we can accomplish this, why bother with a "return" conductor? Such indeed is the logic which permits us to evolve transmission lines into waveguides and cavities.

COMPARISON OF LINES AND GUIDES

If, as suggested, we energize the end of a transmission line by means of a radiating dipole, or loop, rather than by "ohmic" connection to the generator, the transmission line can be excited into the same resonances previously described. Figure 1-25 shows the evolution or transition from transmission line to waveguide. At Fig. 1-25A is shown the use of direct, or ohmic, connections. The next step, at Fig. 1-25B, is injection and monitoring by coupling to the electromagnetic field. The conductors of the transmission line can be extended to form a closed conduit, as in Fig. 1-25C. The longest wavelength (lowest frequency) that can be propagated through such a guide is slightly greater than 2b. Otherwise, similar performance to that of the transmission line is obtained. Finally, at Fig. 1-25D, a coaxial line with inner conductor removed can operate as a waveguide, provided that the high-frequency energy is introduced as a radiated field. Both examples shown operate as shorted quarter-wave sections. Significantly, we see that the conductors of a line can be considered as guides for the lengthwise propagation of the electromagnetic wave. This being the case, we can appreciate that this situation prevails also for a totally enclosed rectangular structure, or for the cylindrical pipe left from removing the central conductor from the coaxial cable. Actually, the study of propagation in such "waveguides" involves complexity not intended in the scope of this book. For our purposes, it is sufficient to realize that quarter-wave and half-wave resonant configurations can exist, which behave similarly to those made up of transmission lines. One of the important differences encountered in waveguides is that, whereas transmission lines will propagate frequencies all the way down to, and including, direct current, every waveguide has a cutoff frequency below which it will not propagate electromagnetic energy. The cutoff frequency is determined by the width, or diameter. For this reason, the use of waveguides is confined to the microwave spectrum, the cross-sectional dimensions at lower frequencies being prohibitive from a practical standpoint.

RESONANT CAVITIES

The simplest resonant cavities are essentially quarter- or half-wave sections of rectangular or round waveguides. In addition to the way in which electromagnetic energy passes down a transmission line, waveguides are generally capable of propagating the energy in different modes, that is, according to different patterns

Fig. 1-25. Transition from transmission line to waveguide involves replacement of ohmic contacts by radiating elements.

of magnetic and electric flux. This sometimes permits undesired resonances; in other instances, it gives rise to physical dimensions that are not so simply related to the resonant frequency as in transmission lines. However, when properly designed, the microwave cavity displays an extremely high Q. If the cavity is made from a material that retains its physical dimensions despite temperature change, and which is mechanically rugged, a frequency-governing oscillating element is obtained that stabilizes the generated microwave frequency to a degree comparable with the stabilization of low frequency oscillators by quartz crystals.

QUARTZ CRYSTALS

Quartz is one of the crystalline substances that is known to exhibit the piezoelectric property. Such crystals undergo a change in physical dimensions when subjected to an electric field; conversely, they generate a voltage when subjected to physical deformation such as might result from application of a pressure or impact (see Fig. 1-26). Slabs or wafers cut from the body of a large crystal retain the piezoelectric property. A wafer of such quartz possesses the mechanical characteristics of mass and compliance. We can therefore expect that a quartz wafer must have a natural oscillation frequency determined by its dimensions. This is, indeed, true. In the simplest and most common case, the natural oscillation frequency (mechanical resonance) is inversely proportional to the thickness of the slab, that is, the thinner the slab, the higher the frequency. Such oscillating elements are commonly called "crystals" although they are, in reality, sections cut from a crystal.

Fig. 1-26. Experiments for demonstrating the piezoelectric effect.

Piezoelectricity and crystallography constitute extensive studies. It will suffice for our purpose to appreciate that the crystal can be mechanically oscillated by the application of electrical signals and, conversely, mechanical oscillation will generate electrical signals. We say that there is a *coupling* between the mechanical and electrical properties.

The Two Resonances in Quartz Crystals

The quartz-crystal oscillating element is capable of behaving either as a parallel- or a series-tuned tank circuit. The equivalent electrical circuit of the crystal oscillating element is shown in Fig. 1-27. This circuit, we see, is a combination series- and parallel tuned network. In Fig. 1-28 the two resonances are illustrated. These are actually very close together, but it is obvious that a crystal oscillator designed to make use of the parallel resonant mode will generate a slightly higher frequency than will be obtained from a circuit in which the frequency is governed by the occurrence of series resonance in the same crystal.

The Relatively Small Tuning Effect of Holder Capacitance

When the piezoelectric vibrations are excited in the parallel-resonant mode of oscillation, we may think of the high inductive reactance of L as being almost, but not completely, cancelled by the high capacitive reactance of C1. The effectively small inductive reactance then is parallel resonated by C2. Whether operating at series or at parallel resonance, the Q of the crystal greatly exceeds that attainable from tank circuits composed of physical inductors and capacitors. It is not possible to construct a coil with the high-Q inductance as represented by the crystal, for either series or parallel resonance.

When the quartz crystal oscillating element is series-resonated it appears as a very large inductance in series with a tiny capacitor and a resistance, this series network then being shunted by the relatively large holder capacitance. The holder capacitance is not part of the equivalent series-resonant circuit and cannot exert a direct tuning influence. The capacitance of the holder, and also other capacitances that might appear in this circuit position, cannot exert a pronounced tuning effect. (Series-connected capacitors produce a resultant capacitance given by the relationship, $(C1 \times C2)/(C1 + C2)$. If C2 is, say, ten times larger than C1, a little mathematical experimentation will show us that C2 can be

Fig. 1-27. Equivalent circuit of a quartz crystal.

doubled with only a slight effect upon the resultant capacitance.)
Although the frictional losses of the vibrating crystal are relatively
high, the very high mass enables attainment of high Q.

Fig. 1-28. The two resonances, f_1 and f_2, of a quartz element.

Conditions for Optimum Stability

Crystal oscillating elements are capable of maintaining frequencies stable from within one part in a million to better than one part in a hundred million, depending very much upon the precautions taken to reduce or compensate the effects of temperature variations. Initially, this precaution is met by detailed attention to the angle at which the wafer is sliced from the body of the crystal. Different cuts have different coefficients of frequency change with respect to temperature. It is possible to cut the wafer in such a way as to have a positive, negative, or zero temperature coefficient. A positive coefficient denotes increase in frequency as the temperature is increased. Most generally, the zero temperature coefficient is considered most desirable. The second method of securing maximum frequency stability from the quartz oscillating element is to operate it at near-constant temperature inside a thermostatically controlled oven. Crystals are commonly associated with low-power oscillators wherein their salient contribution is stabilization of the generated frequency with respect to circuit conditions, temperature, and mechanical shock or vibration. The crystal wafer cannot be excited too vigorously or it will shatter from excessive mechanical forces, or be burned by excessive current that flows through it in the same manner as high-frequency ac flows "through" any capacitor.

A CLOSER LOOK AT CRYSTAL OPERATING CONDITIONS

The resonance diagram (Fig. 1-29) is commonly encountered and suffices to convey a qualitative understanding of the two operating modes of oscillating crystals. However, in practical crystal oscillator circuits, parallel mode operation does not correspond to the anti-resonant frequency of the crystal in which its effective impedance would be at its maximum and would be purely resistive. Instead, the "parallel resonant" crystal operates somewhere in its inductive reactive region. This is why parallel mode crystals are specified to work into a standardized capacitive "load," of 32 pF. On the other hand, crystals intended to operate in the series mode of resonance can be expected to produce their stamped frequency regardless of the reactance placed across the terminals of the crystal.

It is the oscillator circuit that primarily determines whether a crystal will oscillate in its series or parallel resonant mode. However, the manufacturer can process the crystal and its mounting

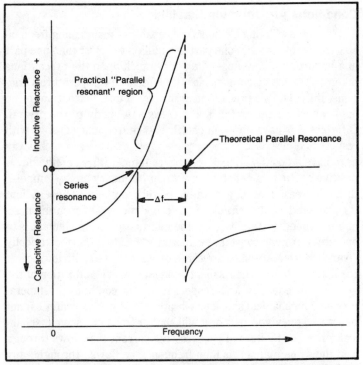

Fig. 1-29. A closer look at the operating conditions of crystals in practical oscillators. Crystals which are spoken of as operating in the "parallel resonance" mode actually oscillate in a frequency region in which they appear as an inductive reactance.

technique so that one or the other mode occurs with optimum results. The important thing to note is that there may be considerable departure from the stamped frequency if a crystal is operated in other than its intended mode. For some applications, this will be of no consequence; for others, the ultimate frequency, especially if derived by multiplying, will be "out of the band."

Δf in Fig. 1-29 has a special significance. It is the difference between the series resonant frequency and the anti-resonant frequency due to holder capacitance alone. The bandwidth of the crystal is known as f and represents the maximum frequency "pulling" theoretically attainable by associating variable series or parallel reactance with the crystal. Other things being equal, high Q crystals are less amenable to frequency pulling than their lower Q counterparts. This statement may have to be somewhat tempered in practice because the lower activity of a low Q crystal may not allow reliable oscillation over the anticipated bandwidth. Also, it must be kept in mind that lower Q means lower frequency stability.

In order to be able to conveniently take advantage of the crystal's bandwidth in applications where it is desirable to vary the frequency, the ratio C2/C1 should be as low as possible. (From Fig. 1-27, C2 is the holder capacitance and C1 is the equivalent series capacitance of the oscillating crystal.) The best the user can do is to specify a low holder capacitance and to minimize circuit and stray capacitances that essentially increase the effective holder capacitance. If these matters are not implemented, an inordinately large tuning capacitor will be needed to change the frequency across the crystal's bandwidth.

Frequency Pulling in Crystal Oscillators

An important use of crystals is found in the VXO, the variable frequency crystal oscillator. Indeed, many "fixed frequency" crystal oscillators are provided with a frequency trimmer. The fact that a crystal oscillator can be made tunable seems a contradiction to the widely-held notion that crystal-stabilized frequencies are "rock-bound." The resolution of this dilemma is in the relatively small amount the frequency of an oscillating crystal can be "pulled." Although small, useful results stem from this phenomenon. For example, such controllable variations of frequency are directly beneficial in amateur communications where even a slight increase or decrease in carrier frequency can lessen the interference of other stations. In receivers, the ability to fine-tune beat-frequency oscillators and "fixed" local oscillators helps optimize reception. The fact that the frequency of a crystal can only be pulled a small percentage become less restricting when crystal frequencies are multiplied to vhf and uhf bands. This also facilitates frequency-modulation inasmuch as the deviation increases along with the frequency multiplication factor; thus 500 Hz deviation at 14 MHz becomes 5 kHz deviation at 140 MHz. And frequency pulling has long been used for frequency standard and instrument time-bases where a small variable-capacitor, or sometimes an inductor, has been used to set the crystal frequency.

The practical aspects of frequency pulling remain an art as well as a science. Not all crystals are equally amenable to having their "natural" frequency varied. Much depends upon the cut of the crystal, its activity, its holder capacitance, and several elusive factors. Sometimes an otherwise "rubbery" crystal causes trouble with spurious frequencies if it is pulled too far. Obviously, the oscillator circuit and its practical implementation are very much involved. If stray capacitance or inductance is needlessly high, the pulling range tends to be narrowed. And if strong positive feedback is not available, the crystal oscillator will not start when "detuned."

Crystals that oscillate in their series-resonant mode can have their frequency increased by the insertion of a small series-connected variable capacitor. A typical example of this is seen in the Clapp version of the Colpitts oscillator. Parallel-resonant crystals can have their frequencies lowered by means of a small shunt-connected variable capacitor. Also, crystals oscillating in their series-resonant mode can have their frequencies *lowered* by small series-connected inductors. Generally speaking, overtone crystals are not good candidates for frequency pulling. An exception, however, is when the overtone crystal is deliberately used in its fundamental-frequency mode. In this instance there appears to be evidence that it is capable of exceptionally-wide frequency pulling.

The crystal frequency pulling phenomenon lends itself very advantageously to varactor tuning, a technique particularly suitable for remote tuning and for frequency modulation with an audio signal. The varactor is essentially a reverse-biased pn diode which provides a useful range of capacitance in response to varying dc bias voltage. One of its attributes for crystal frequency-pulling is the inordinately-small minimum capacitance available.

Some Practical Considerations of Oscillating Crystals

Older-type crystals (those commonly found on the surplus market) were physically larger than their more modern counterparts, used natural quartz, and a pressure-mounting technique in a gasket-sealed package. These were capable of safe operation at higher excitation currents than later crystals and tended to be more amenable to frequency pulling than are modern solder-seal types using synthetic quartz. However, the other side of the coin is that modern crystals are more reliable. They are less likely to be plagued by aging effects, by decline in activity, or by spurious emissions. At the same time, their temperature coefficient is either smaller or more predictable. By appropriate selection of a modern crystal and its oscillating circuit, more than enough frequency pulling is attainable for most purposes. At best, frequency pulling is a vernier effect and cannot compete with the VFO where wide tuning ranges are needed. The frequency-pulling range for many practical applications is limited to about 1/500 of the minimal crystal oscillation frequency. Attempts at greater pulling may seriously sacrifice frequency stability.

Certain oscillator circuits, such as the Pierce oscillator and some of those used with logic ICs, will produce self-excited oscillations without the crystal being inserted in its socket. When frequency pulling is used with such circuits, attempts at excessive pulling may result in an abrupt jump to the self-excited frequency. On the other hand, circuits such as the conventional Colpitts will simply go dead if the demanded frequency pulling is too great. Even here a word of caution is in order: the Colpitts circuit is generally a ready oscillator for either series-resonant or parallel-resonant crystals. Thus, it can happen that excessive frequency pulling can cause an abrupt transition between these modes with an attendant discontinuity in the oscillation frequency. This is not likely to happen, however, unless a crystal intended for parallel resonance is operated in its series-resonant mode, or vice versa.

Crystals intended for crystal filters are less active than oscillator crystals and are generally not good candidates for frequency-pulling applications. This is unfortunate inasmuch as these crystals have the otherwise desirable feature that they are remarkedly free of spurious frequency tendencies. Finally, the manufacturer's data pertaining to temperature coefficient, drift, and Q pertain to "on-frequency" operation. For demanding applications involving frequency pulling, it is probably best to consult the manufacturer who has tricks of the trade to optimize frequency-pulling ability, as well as other crystal parameters.

From a mechanical standpoint, there are several ways in which a crystal can vibrate in order to manifest itself as a piezoelectrical resonator. The flexural mode of vibration is employed for very-low frequency crystals, enabling operation down to 200 Hz. This vibrational mode is not likely to be encountered beyond 100 kHz where one begins to find competition between the extensional and face-shear vibrational modes. From one megahertz to the vicinity of 20 to 35 MHz the thickness-shear mode of vibration is used and the AT and BT cuts predominate. The AT cut is significant in that it generally yields the best temperature characteristics. Finally, overtone crystals also emply AT and BT cuts, and also vibrate in the thickness-shear mode. However, the overtone crystals are processed and mounted in such a way that odd overtones of the fundamental frequency are readily excited in appropriate circuits. Such crystals are practical to the ninth overtone, and may yield frequencies up to about 250 MHz.

The above information is a miniscule condensation of a voluminous subject. Fortunately, one does not have to be versed in crystallography to design and use crystal oscillators. At the outset, however, it should be known whether the oscillator circuit makes use of series or parallel resonance in the crystal. Although most crystals, regardless of cut or vibrational mode can be operated in either the series or the parallel resonant condition, the best use of a crystal stems from operation in the resonant mode prescribed by the manufacturer. Such intended operation usually produces the highest Q, the best stability, the easiest starting, the most favorable temperature-coefficient, and the closest operation to a stipulated frequency.

Some circuits obviously require a certain type of resonance. For example, the Butler and Pierce oscillators need crystals optimized for their series-resonance mode. The Miller oscillator (otherwise known as tuned-plate, tuned grid; or tuned collector, tuned base; or tuned drain, tuned gate) has to have parallel-resonant crystals. In these instances, it is customary for the manufacturer to stipulate a frequency corresponding to parallel resonance with a crystal "load" of 32 pF.

In other circuits, the mode of resonance may not be obvious. Nor can one infer series resonance just because a small capacitor is connected in series with the crystal. For example, the ordinary Colpitts oscillator will cause its crystal to resonate in its parallel mode providing the crystal is made to develop its highest activity in this mode. If, however, the crystal has been optimized to perform in the series resonant mode, the Colpitts oscillator will become a Clapp oscillator (a Colpitts with a series-resonant tank) even though this was not the designer's intent, and even though it is not apparent from the topography of the circuit.

Controlling and Optimizing the Temperature Dependency of Quartz Crystals

The effective Q of an oscillating quartz crystal may be made very high by appropriate processing and manufacturing techniques and its activity can be made great enough so that reliable start-up is attainable without inordinate emphasis on the active device or on loading. But this is not enough. The temperature dependency of the crystal must be given due consideration during the processing stage and also during circuit design. For best stability, both factors are important. Crystallography is a voluminous subject and will not be dealt with in this book, but the temperature behavior of the popular AT-cut crystal provides good insight into the situation. This is especially true when it is recognized that other crystal cuts generally have worse temperature behavior than the AT-cut.

Figure 1-30 shows several of the numerous cuts which may be processed from the quartz crystal. The cut greatly affects temperature coefficient, vibrational mode, Q, activity, spurious responses, harmonic behavior, and frequency capability. Aside from the use of different angular cuts, oscillating crystals can assume other formats. One popular one for the approximately 32 kHz oscillator in electronic watches makes use of a quartz element in the shape of a tuning fork. The AT-cut is at a nominal 35 degrees with respect to the Z axis of the quartz crystal. However, slight angular deviations of the cut enables the manufacturer to "fine-tune" the all-important temperature coefficient. Figure 1-31 shows how this comes about.

Not only can a substantially-zero temperature coefficient be achieved over a useful temperature range, but it is possible to design in a temperature coefficient that will then cancel the temperature coefficient of the rest of the oscillator circuit. In this way, the overall temperature coefficient of a crystal oscillator can be made very small.

Two other ways are useful in combating the temperature dependency of crystals. Circuit components, usually capacitors, are

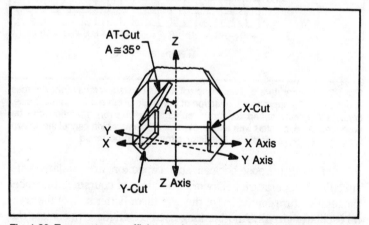

Fig. 1-30. Temperature-coefficient and other characteristics depend upon the angle of cut. The AT-cut is processed approximately at 35 degrees with respect to the Z axis. Minor angular deviations of this cut enables "fine-tuning" of the temperature coefficient. The X-cut is made perpendicular to the X axis. The Y-cut is made perpendicular to the Y axis. Many other cuts have been used, each with its unique parameters and performance trade-offs.

Fig. 1-31. Temperature characteristics of AT-cut crystals versus slight changes in cut-angle. By appropriate selection of the angle of cut relative to the Z axis, crystals optimally suited for room temperature or oven operation can be produced. Also, crystals with temperature coefficients which cancel the known temperature behavior of the oscillator circuitry can be made.

useful for this purpose because such capacitors are available with specified temperature behavior. The idea, of course, is to cancel the temperature coefficient of the crystal, or better still, of the crystal oscillator circuit. Another long-used technique is to enclose the crystal in a temperature-stabilized oven. To be useful, the oven temperature must be above the ambient temperature. Ovens that cycle on and off have been much used, but proportional types give more refined results. Sometimes it is expedient to include other oscillator components within the oven and in certain cases, the entire crystal oscillator circuit may be oven stabilized.

After any or all of the alluded safeguards against temperature dependency are incorporated, several design considerations are still needed for attainment of best frequency stability. The crystal should not be driven too hard and should be "decoupled" as much as possible from the active device. Finally, a high-isolation buffer amplifier is needed to prevent the load from affecting the oscillation frequency.

The Dielectric Resonator

It has long been known that the dimensions of a microwave resonant cavity could be dramatically reduced by filling it with a dielectric or insulating material. Indeed, a block or slab, or other geometric shape of such material can demonstrate resonant behavior without being encased in metal. However, materials available until recently have been characterized by high loss, physical instability, and by high temperature coefficients. Previously, crystalline rutile and strontium titanate were the best dielectric materials available for microwave and millimeter-wave dielectric resonators. These materials exhibited temperature coefficients of 1000 parts per million per °C, and greater. Now available are certain barium tetratitanate materials that show temperature coefficients in the vicinity of -50 at 10 GHz. The dielectric constant of such material is 38. This is an important factor inasmuch as the size of a dielectric resonator is inversely proportional to the square root of the material's dielectric constant.

The resonant frequency of a dielectric resonator is a function of its dielectric constant, its geometry, and its boundary conditions. This brings us to an interesting aspect of the newer materials developed for dielectric resonators. It turns out that temperature induced variations in dielectric constant and in physical dimensions are such that they nearly cancel. That is why very low temperature coefficients of frequency change is attainable. At the same time, dielectric losses in the new materials are so low that unloaded Q's of 6000 and better are attainable. Although the dielectric resonator cannot yet be said to compare well with quartz crystals at much lower frequencies, it must be remembered that materials technology and circuit techniques are much more difficult at microwave and millimeter-wave frequencies than at the more familiar radio frequencies.

The functional diagrams (Fig. 1-32) show the use of the die-

Fig. 1-32. Deployment of the dielectric resonator in two types of microwave oscillators. (A) Negative resistance oscillator. (B) Feedback oscillator.

lectric oscillator in microwave oscillators. The negative resistance oscillator (A) may be a Gunn or IMPATT diode, whereas the active element (B) is likely to be a gallium-arsenide field-effect transistor. In both cases, the dielectric resonator behaves as a narrowband filter that is coupled to the microwave circuitry as depicted in Fig. 1-33.

MAGNETOSTRICTIVE ELEMENT

The piezoelectric characteristic of the quartz crystal is not the only medium whereby the mechanical and electrical properties of a vibrating element can be coupled so that mechanical movement generates an electrical signal and vice versa. A somewhat analogous situation is found in the magnetostrictive effect in certain magnetic materials, especially in nickel alloys. Magnetostriction is the change in length produced in rods of such materials in response to a magnetic field, and the converse change in magnetization that results when such rods are subjected to mechanical tension or compression. The word, "magnetostriction," is misleading in that it suggests a shortening of length only. Actually, both shortening and lengthening exist in different alloys, and in a few instances, in the

Fig. 1-33. Coupling mechanism and filtering action of a dielectric resonator. The rf magnetic fields provide energy linkage into and out of the dielectric resonator, which behaves as a selective filter.

same alloy (see Fig. 1-34). With most magnetostrictive alloys the shortening (or lengthening) will occur for each half of the ac cycle. This means that the rod undergoes two complete cycles of mechanical vibration for each cycle of the exciting force. (This did not happen with the piezoelectric crystal because the crystal becomes thicker for one half-cycle of applied potential, then thinner for the second half-cycle. In other words, the vibrations of the crystal "follow" the electrical oscillations.) It will be shown that the magnetostrictive element is suggestive of a transformer with very loose

Fig. 1-34. Fractional change in length caused by magnetostriction effect in several materials.

45

coupling at off-resonant frequencies, but with tight coupling at resonance.

Need for Bias

The frequency-doubling property of the vibrating magnetostrictive rod is generally not desirable. In order to make the rod maintain pace with the applied magnetic oscillation, it is necessary to superimpose a constant magnetic force. This is readily obtained from the effect of a direct current that is either applied to a separate bias winding, or allowed to circulate in a signal winding. This magnetic bias prevents reversal of the sense of magnetization, thereby causing the rod to vibrate at the same frequency as the signal (see Fig. 1-35). An analogous situation is found in the earphone, which must be provided with a small permanent magnet so that the magnetic force acting on the diaphragm does not reverse its sense. Without such provision, severe distortion would occur because the diaphragm would tend to vibrate at a frequency twice that of the electromagnetic signal representing the speech or music.

Frequency/Length
Relationships in Magnetostrictive Element

A nickel-alloy rod about four inches in length vibrates at a frequency of 25 kilohertz. The frequency of resonant vibration is inversely proportional to length so that practical limitations are encountered at both low and high frequencies. We see that a one hundred kilohertz magnetostrictive element is on the order of an inch in length, whereas a twenty-inch rod is required for five-kHz oscillation.

THE MAGNETOSTRICTION OSCILLATOR

When connected in conjunction with a suitable oscillation-provoking device, such as a vacuum tube, an ac signal applied to an exciting coil subjects the biased rod to a magnetic force of cyclically varying strength. This causes the rod to change its length accordingly. The changes in rod length are accompanied by changes in its magnetization that in turn induce a signal in another coil. The induced signal is of maximum amplitude at the frequency corresponding to vibrational resonance of the rod, for it is then that the changes in length and magnetization are greatest. The signal undergoes amplification in the vacuum tube, or other device, then is

Fig. 1-35. The magnetostrictive oscillating element.

reapplied to the exciting coil of the magnetostrictive oscillating element. In this way, oscillation is maintained at the mechanical frequency of resonance in the rod. When the magnetostriction element is associated with an amplifier, phasing of the windings is made opposite to that necessary for conventional "tickler" or Hartley feedback oscillators. This is true because induced voltage from magnetostriction is, at any instant, of opposite polarity to voltage generated by ordinary transformer action. This is fortunate, for otherwise it would be very difficult to prevent oscillation at an undesired frequency due to electromagnetic feedback.

THE TUNING FORK

Besides the quartz crystal and the magnetostrictive rod, a third electromechanical oscillating element is of considerable importance. This is the electrically driven tuning fork. A steel bar bent into a U shape can be excited into vibration by means of a solenoid placed near one of the fork ends. As we might expect, the maximum intensity of the vibration occurs when the frequency of the solenoid current is the natural resonance frequency of the fork. The region near the U junction is a nodal region where vibrational amplitude is at a minimum. On the other hand, the ends of the fork prongs execute transverse vibrations of relatively great amplitude. Suppose we place a second solenoid at a corresponding position near the opposite prong. If, then, we monitored the induced voltage in this winding, we would find that it peaked up sharply at the resonant frequency of the fork (see Fig. 1-36). The flux linking the pickup winding undergoes its most intense variation at resonance due to the mechanical motion of the fork. This, in turn, produces the greatest change in flux in the pickup winding thereby causing the observed rise in ac voltage developed in the pickup winding. When the fork is connected to an amplifier to form an oscillating circuit, the pickup winding delivers a signal to the input of the amplifier. The amplified signal appearing at the output of the amplifier energizes the exciting solenoid. The resultant magnetic impetus then serves to regenerate this oscillation cycle and sustained oscillation ensues.

Applications

Over the frequency range of about 100 Hz to several tens of kilohertz, such a fork behaves as a high-Q energy storage system capable of providing excellent frequency stability. Other types of tuning forks have been used. In one type, the fork carries contacts and the device becomes essentially a buzzer. In another version, the pickup signal is obtained through the use of a carbon microphone.

Fig. 1-36. The tuning fork.

Fig. 1-37. RC phase-shift network.

RC NETWORKS AS OSCILLATING ELEMENTS

Numbered among oscillating elements are certain combinations of resistance and capacity. Such networks are not "natural" supporters of oscillations; rather, they become oscillating elements in special amplifier circuits. Even in those RC networks in which sharp frequency selectivity exists, the tendency to "ring" that characterizes LC tank circuits is entirely absent. (However, ringing can be demonstrated with the network connected to its feedback amplifier.) The RC network, no matter what its form, functions essentially as a phase-shifting circuit. Generally, the oscillator circuit operates at the frequency at which the RC network produces a 180° phase shift or zero, whichever corresponds to positive feedback. Other frequencies produce phase shifts that correspond to negative feedback. As a consequence there is high discrimination against such frequencies. This permits generation of a very nearly pure sine wave. To accomplish the same thing with an LC tank, the Q would have to be quite high.

One of the useful properties of the RC networks is that frequency is inversely proportional to capacitance, rather than to the square root of capacitance as in an LC tank circuit. This means that a given change in capacitance produces a much greater frequency change than in an LC tank circuit. If we increase capacitance nine times in the LC tank, we decrease the resonant frequency by one-third. The same increase in capacitance in RC networks decreases frequency by one-ninth. It is, however, often necessary to simultaneously tune two or three elements in the RC network. An additional disadvantage is the fact that the output power of the RC oscillator is necessarily low by virtue of the dissipation in the resistive elements.

Examples of RC Networks Suitable as Oscillating Elements

One of the simplest means of obtaining phase inversion (180°

49

phase shift) is to cascade three similar RC high-pass filter circuits. The resulting phase-shift network is shown in Fig. 1-37. The simple prototype RC high-pass network is limited in its phase-shifting capability. Extended phase shifting is obtained by cascading two or more of the prototype circuits. When such a network is connected between input and output of a single-tube amplifier, oscillation occurs at the frequency at which the network compensates the natural 180° phase shift in the amplifier. For three cascaded RC pairs as shown, 180° phase shift occurs at a frequency approximately equal to:

$$f = \frac{1 \times 10^6}{2\pi\sqrt{6}\,RC}$$

where,

f is the frequency in hertz,

R is the resistance in ohms,

C is the capacitance in microfarads.

A similar phase inversion is produced at the "resonant" frequency of the parallel-T bridge illustrated in Fig. 1-38. For this bridge, resonant frequency occurs at:

$$f_0 = \frac{1}{2\pi\,(C_1)\,(R_2)}$$
$$= \frac{1}{2\pi\,(C_2)\,(R_1)}$$

Fig. 1-38. The parallel-T network.

Fig. 1-39. The Wien bridge.

where,

    f_0 is the frequency in hertz,

    R_1 and R_2 are in ohms,

    C_1 and C_2 are in farads,

    R_2 equals 2 R_1,

    C_2 equals 2 C_1.

In this case, the phase inversion is accompanied by maximum attenuation of the signal traversing the network. This is not the desirable response characteristic because it tends to defeat our purpose if we are to invert the output signal of an amplifier, but at the same time drastically reduce its amplitude. However, it is comparatively easy to compensate for the high attenuation by providing sufficient voltage amplification in the amplifier associated with the network. When this is done, oscillation takes place at the frequency "null" of the network. Another bridge with phase and amplitude response similar to that of the parallel-T network is the Wien bridge depicted in Fig. 1-39. This basic bridge is not readily applied to RC oscillators because of the four-terminal arrangement. It is shown merely for academic interest. It is, however profitable to employ this bridge in a somewhat different manner, using only the two reactive arms in the positive-feedback path of the oscillator. When so utilized, it displays the phase and amplitude characteristics shown in Fig. 1-40. Actual oscillators using these RC networks are described in detail in the section dealing with the theory of oscillation.

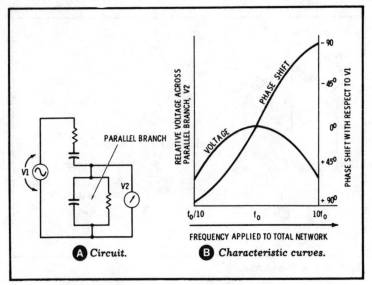

A *Circuit.*   **B** *Characteristic curves.*

Fig. 1-40. Characteristics of reactive arms of Wien bridge.

LC NETWORKS AS PHASE SHIFTERS

Of utmost importance are the phase properties of certain LC networks. These LC networks produce phase inversion at a discrete frequency, thereby enabling a portion of the output signal of an oscillator tube to be returned to the input circuit in proper phase relationship to reinforce the signal already there. The lC networks illustrated in Figs. 1-41 and 1-42 are basic to the operation of the majority of LC feedback oscillators. However, they are not always immediately recognizable as being equivalent to these simple circuits. For example, what we have shown as an inductor may actu-

$$f_0 = \frac{1}{2\pi \sqrt{(L1 + L2) C}}$$

$$e_2 = -\frac{L2}{L1} e_1 \text{ (AT RESONANCE)}$$

A *Circuit.*   **B** *Voltage waveform and formulas.*

Fig. 1-41. An LC network that produces phase inversion (Hartley type).

Fig. 1-42. Additional LC networks that produce phase inversion.

ally be a parallel-tuned tank operated slightly below its resonant frequency, under which condition such a tank appears inductive. Conversely, the inductor may be a series-tuned tank operated slightly above its resonant frequency. Instead of physical inductors or inductor-capacitor combinations, an actual oscillator may employ a piezoelectric crystal. In the chapter on practical oscillators, we will find these LC networks directly relevant to the basic operating principles of important oscillator circuits. For instance, the network of Fig. 1-41 underlies the theory of oscillation in the Hartley and also in the tuned-plate tuned-grid oscillator. Similarly, the principle of the well-known Colpitts oscillator depends upon the phase-inverting property of the LC network shown in Fig. 1-42A.

Chapter 2

Active Devices of Oscillators

A N ACTIVE, OR OSCILLATION-PROVOKING, DEVICE IS AS ESSEN-
tial to the oscillator as are frequency-determining elements.
Such devices remove the effect of dissipative losses that are always
present in frequency-determining elements at ordinary tempera-
tures. Interestingly, we will encounter diverse methods for accomplishing this basic objective. We will investigate switching devices,
amplifying elements, the use of "negative resistance," and the excitation of oscillation by sparks and electron beams. In so doing,
we will find oscillators in which the frequency-determining elements
and the active device are physically separate; conversely, in other
oscillators these two sections will be found associated as a single
integral unit.

SWITCHING ACTIONS IN TUBES

The vacuum tube, in addition to its traditional role as a proportionate amplifier, is a very effective switch for supplying exciting
pulses to LC circuits and their counterparts. When so used, the action is that of an "amplifying switch," one in which the switching
of oscillatory power can be provoked by switching cycles of considerably lower power-level. In the triode vacuum tube, two more
or less opposite switching actions are important. On one hand, plate-
current cutoff is produced when the grid bias becomes sufficiently
negative; on the other hand, plate-current saturation occurs when
the grid is made sufficiently positive (see Fig. 2-1). The pulses

A *Typical plate-current/grid-voltage relationship.*

B *Negative grid.*

C *Positive grid.*

Fig. 2-1. Vacuum tube can operate as switching element when suitable grid bias is applied.

formed by one, the other, or both mechanisms are very suitable for provoking oscillations. For most oscillator applications, a continuous train of pulses is generated at a pulse rate equal to the frequency of the frequency-determining elements (generally a resonant LC circuit). In some special cases, the switching pulse rate can be an integral fraction of the oscillation frequency.

The transistor behaves somewhat similarly to the vacuum tube when employed as a switching or pulsing element. Npn and pnp transistors perform similar circuit functions, but the two types require oppositely polarized electrode supplies (Fig. 2-2). Notice that the npn transistor is more reminiscent of the vacuum tube in the matter of supply voltages. In the npn transistor and in the vacuum tube, the "collector" is polarized positively with respect to the "emitter." In both instances output current is increased by making the control element more positive. Indeed, the operation of such

55

Fig. 2-2. Output diode section of transistor acts as switch as bias current is applied or removed from input diode section.

a transistor is, for many purposes, quite closely simulated by the performance of a triode supplied with positive grid bias. We can make the simulation even closer by postulating the use of a zero-bias triode, one that does not permit appreciable plate current without application of positive grid bias. It is often stated that the tube is a voltage-actuated device, whereas the transistor is current-actuated. This distinction holds true only when the tube is operated with negative grid bias. On the other hand, collector-current cutoff of a transistor occurs at zero input current.

AMPLIFICATION

Amplification is one of the very important processes associated with most oscillators. The vacuum tube and the transistors are the most widely used amplifying elements. Now would be an appropriate time in our study of oscillators to review some of the amplifying characteristics of these elements that are particularly relevant to oscillators. The ability of the tube to faithfully reproduce weak speech and music signals at stronger power levels derives from its function as a proportional amplifier (Fig. 2-3). By this, we mean that the output signal can be proportional to the input signal. If the input signal doubles in amplitude from 0.1 volt to 0.2 volt, the amplified output signal also doubles in amplitude, say, from 10 volts to 20 volts.

The extreme opposite of proportional amplification is seen in the operation of a switching device. For example, the electromagnetically actuated relay or contactor provides high power amplification, but there is no relationship between the voltage or current supplied to the solenoid and the voltage or current switched by the contacts. Between the extreme or proportional amplification and the switching process, there are output-input relationships that are best described as being nonlinear. Generally this means that output will follow the amplitude variations of the input or actuating signal, but not in a proportional manner. In the nonlinear amplifier, the input signal may increase from 0.1 volt to 0.2 volt; in response, the amplified output signal may increase from 10 volts to 15 volts, or perhaps from 10 volts to 35 volts. A significant point

Fig. 2-3. Linear (proportional) amplification in Class-A amplifier.

with respect to these matters is that a linear amplifier is linear only over a limited range of signal amplitude. Any linear amplifier can be driven into its nonlinear region by sufficient signal amplitude.

Class-A Amplification

Linear operation of vacuum-tube amplifiers is obtained by Class-A operation. Such operation results when the dc grid and the ac signal are such that plate current flows throughout the complete cycle of the input signal. Though not generally stipulated, Class-A operation is generally understood to denote proportional amplification. For example, the plate current should not be impeded by approaching saturation in its attempt to follow the excursions of the input signal. Figure 2-4 shows low-frequency properties of Class-A amplifiers.

Class-B Amplification

The Class-A amplifying mode is not the only, nor necessarily the most desirable, way in which a tube can amplify. For example, if the grid bias and the signal are of such amplitudes that the plate current of the tube is cut off during the negative half of the input cycle, we have what is known as a Class-B amplifier. In a sense, we are rectifying the input signal and amplifying only the resultant unidirectional wave. In order to accomplish this, the grid is usually driven positive for part of the positive half of the input voltage cycle. Because of the resultant grid current, the Class-B amplifier must be supplied with considerable input power. However, the degradation of efficiency and power-multiplying capability is more than compensated by the fact that plate current is due entirely to response to the input signal. We observe in Fig. 2-5 that this is not so in the Class-A amplifier wherein high plate current exists in the absence of any input signal. Thus, for a given power-handling capability, the Class-B amplifier is not called upon to dissipate needless (from the standpoint of signal reproduction) dc plate power. In audio work, the single Class-B amplifier is not suitable because it provides response for only half of the signal waveform. This shortcoming is overcome by employing two such amplifiers in push-pull wherein one tube delivers output proportional to one half-cycle of input signal, and the other tube delivers power proportional to the opposite half-cycle of input signal. The original waveform of the input signal is then restored by appropriate combination of the two amplified half-cycles in the load, usually the speaker.

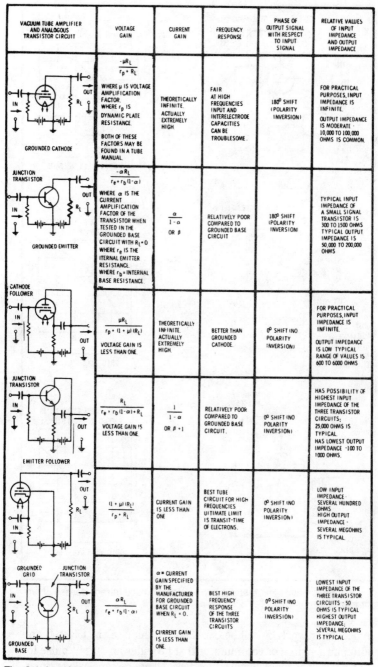

VACUUM TUBE AMPLIFIER AND ANALOGOUS TRANSISTOR CIRCUIT	VOLTAGE GAIN	CURRENT GAIN	FREQUENCY RESPONSE	PHASE OF OUTPUT SIGNAL WITH RESPECT TO INPUT SIGNAL	RELATIVE VALUES OF INPUT IMPEDANCE AND OUTPUT IMPEDANCE
GROUNDED CATHODE	$\dfrac{-\mu R_L}{r_p + R_L}$ WHERE μ IS VOLTAGE AMPLIFICATION FACTOR. WHERE r_p IS DYNAMIC PLATE RESISTANCE. BOTH OF THESE FACTORS MAY BE FOUND IN A TUBE MANUAL.	THEORETICALLY INFINITE. ACTUALLY EXTREMELY HIGH	FAIR AT HIGH FREQUENCIES INPUT AND INTERELECTRODE CAPACITIES CAN BE TROUBLESOME.	$180°$ SHIFT (POLARITY INVERSION)	FOR PRACTICAL PURPOSES, INPUT IMPEDANCE IS INFINITE. OUTPUT IMPEDANCE IS MODERATE · 10,000 TO 100,000 OHMS IS COMMON.
JUNCTION TRANSISTOR **GROUNDED EMITTER**	$\dfrac{-\alpha R_L}{r_e + r_b(1-\alpha)}$ WHERE α IS THE CURRENT AMPLIFICATION FACTOR OF THE TRANSISTOR WHEN TESTED IN THE GROUNDED BASE CIRCUIT WITH $R_L = 0$. WHERE r_e IS THE INTERNAL EMITTER RESISTANCE. WHERE r_b = INTERNAL BASE RESISTANCE.	$\dfrac{\alpha}{1-\alpha}$ OR β	RELATIVELY POOR COMPARED TO GROUNDED BASE CIRCUIT	$180°$ SHIFT (POLARITY INVERSION)	TYPICAL INPUT IMPEDANCE OF A SMALL SIGNAL TRANSISTOR IS 300 TO 1500 OHMS TYPICAL OUTPUT IMPEDANCE IS 50,000 TO 200,000 OHMS
CATHODE FOLLOWER	$\dfrac{\mu R_L}{r_p + (1+\mu)(R_L)}$ VOLTAGE GAIN IS LESS THAN ONE.	THEORETICALLY INFINITE. ACTUALLY EXTREMELY HIGH.	BETTER THAN GROUNDED CATHODE.	$0°$ SHIFT (NO POLARITY INVERSION)	FOR PRACTICAL PURPOSES, INPUT IMPEDANCE IS INFINITE. OUTPUT IMPEDANCE IS LOW TYPICAL RANGE OF VALUES IS 600 TO 6000 OHMS
JUNCTION TRANSISTOR **EMITTER FOLLOWER**	$\dfrac{R_L}{r_e + r_b(1-\alpha) + R_L}$ VOLTAGE GAIN IS LESS THAN ONE.	$\dfrac{1}{1-\alpha}$ OR $\beta + 1$	RELATIVELY POOR COMPARED TO GROUNDED BASE CIRCUIT.	$0°$ SHIFT (NO POLARITY INVERSION)	HAS POSSIBILITY OF HIGHEST INPUT IMPEDANCE OF THE THREE TRANSISTOR CIRCUITS; 25,000 OHMS IS TYPICAL. HAS LOWEST OUTPUT IMPEDANCE -100 TO 1000 OHMS.
	$\dfrac{(1+\mu)(R_L)}{r_p + R_L}$	CURRENT GAIN IS LESS THAN ONE	BEST TUBE CIRCUIT FOR HIGH FREQUENCIES ULTIMATE LIMIT IS TRANSIT-TIME OF ELECTRONS.	$0°$ SHIFT (NO POLARITY INVERSION)	LOW INPUT IMPEDANCE - SEVERAL HUNDRED OHMS HIGH OUTPUT IMPEDANCE - SEVERAL MEGOHMS IS TYPICAL.
GROUNDED GRID **JUNCTION TRANSISTOR** **GROUNDED BASE**	$\dfrac{\alpha R_L}{r_e + r_b(1-\alpha)}$	α = CURRENT GAIN SPECIFIED BY THE MANUFACTURER FOR GROUNDED BASE CIRCUIT WHEN $R_L = 0$. CURRENT GAIN IS LESS THAN ONE.	BEST HIGH FREQUENCY RESPONSE OF THE THREE TRANSISTOR CIRCUITS	$0°$ SHIFT (NO POLARITY INVERSION)	LOWEST INPUT IMPEDANCE OF THE THREE TRANSISTOR CIRCUITS · 50 OHMS IS TYPICAL HIGHEST OUTPUT IMPEDANCE; SEVERAL MEGOHMS IS TYPICAL.

Fig. 2-4. Low-frequency properties of Class-A amplifiers.

Fig. 2-5. Schematics and response characteristics of different classes of amplifiers.

In oscillator work, it will be beneficial to think of the Class-B amplifier as a switch or commutator that provides an output pulse for each cycle of ac input signal. Here we shall not be concerned primarily with fidelity of reproduction, for the output waveform will be governed by the properties of tank circuits and will, ideally, be sinusoidal.

Extension of Class-B Concept

In oscillators employing amplifiers, the amplification need be high only at the single frequency of oscillation, as determined by resonant circuits. As mentioned, fidelity of reproduction of the amplifier input signal is not important because a high-Q resonant tank in the output circuit of the amplifier will restore or convert the output wave to a sinusoid. This is desirable because the sine wave alone is free from harmonic content. All other waveforms are, in reality, pulses, and contain varying degrees of harmonic energy. We found that the Class-B amplifier, with grid biased at plate-current cutoff, performed more efficiently for single-frequency operation than did the Class-A amplifier. It is only natural to wonder whether this technique can be extended further with even greater gain in efficiency. Why not bias the grid with a higher negative voltage than is needed for plate-current cutoff? In order to maintain operation with such a highly biased amplifier, we would, of course, need to increase the amplitude of the input signal. Tube conduction would occur only during the top portion of the input wave, rather than during the entire positive excursion of the input wave as in the Class-B amplifier. The output wave would therefore consist of relatively narrow pulses. We might suspect however, that such pulses would be entirely adequate to shock-excite an LC tank with resonant frequency equal to the output-pulse repetition rate. We know that if the Q of such a tank were high enough, the resultant oscillations developed across it would suffer very little damping between plate-current pulses.

THE CLASS-C AMPLIFIER

The hypothetical operating condition described is actually that of the Class-C amplifier. This amplifier is biased with approximately twice the value of negative grid bias needed to produce plate-current cutoff (Fig. 2-6). It is, indeed, more efficient for single frequency operation than is the Class-B amplifier. When the resonant frequency of the parallel-tuned tank circuit is the same as the input frequency, the unidirectional plate-current pulses are offered a high-impedance path by the tank circuit. This, in turn, results in two very important operating features: first, the dc plate current is at

$$f_0 = \frac{1}{2\pi\sqrt{LC}}$$

WHEN $f_0 = f_{IN}$

WHEN f_0 AND f_{IN} ARE CONSIDERABLY DIFFERENT FREQUENCIES.

* TWICE THE VALUE OF NEGATIVE BIAS REQUIRED TO CUT OFF PLATE CURRENT.

A *General circuit conditions.*

B *Dc plate current as a function of tank tuning.*

Fig. 2-6. Class-C amplifier with resonant tank in output circuit.

a minimum (Fig. 2-6). For other frequencies of the tank circuit, that is, when the output tank is not resonated at the same frequency as the amplifier input signal, the dc plate current is higher. Secondly, it follows that the signal-frequency voltage developed across the tank circuit must be at its maximum value when the input signal frequency and the output tank resonant frequency are identical. In other words, we obtain maximum output with minimum dc plate current. Still another way of viewing this situation is to recognize that, because of the high impedance of the output tank circuit, the instantaneous plate current is minimum when the instantaneous plate voltage is maximum, and vice versa. Inasmuch as the power dissipated at the plate of the tube is the product of plate current and plate voltage, we now see that the Class-C amplifier delivers maximum output from its plate tank circuit under the operating condition of relatively low plate-power dissipation. The plate-current pulses are much narrower, that is, their duty cycle is lower than in the case of the Class-B amplifier. Accordingly, plate dissipation is lower and efficiency is greater than in the Class-B amplifier.

Action of Tank Circuit in Class-C Amplifiers

It is important that we gain a clear insight into the reason the unidirectional plate pulses are offered high impedance by the tank circuit. We understand that the tank is resonated at the frequency that is the same as the pulse-repetition rate of the plate-current pulses. We recall from our knowledge of pulses that such a periodic pulse-train is made up of a dc component, a fundamental frequency, and numerous harmonic frequencies that are whole-number multiplies of the fundamental frequency. The resonant tank circuit limits the current at the fundamental frequency only. Most of the energy of the pulse-train is contained in the fundamental frequency. Therefore, the limitation of current at the fundamental frequency produces a pronounced dip in plate current, for only the dc component and the harmonics "see" a low-impedance path through the tank circuit. The dc component is actually due to the unique phase combination of the harmonics, and is depleted in amplitude when the first harmonic (fundamental) is "strained out" from the total plate current through the tank. The result—a low reading on the dc plate-current meter.

A very large part of the oscillator family makes use of "C" amplifiers in conjunction with parallel-tuned tank circuits. It is of the utmost importance that the essential operating features of Class-C amplifiers (as here expanded) be clearly comprehended. A basic qualitative understanding of the Fourier theorem as it relates to pulse composition is indispensable. It is strongly recommended that this subject be reviewed if it is felt that appreciation is lacking of the main features of nonsinusoidal waves.

Action of Grid Current in Class-C Amplifiers

Because of the fact that the grid consumes current from the signal source only when the positive excursion of the signal wave attains amplitude sufficient to overcome the negative dc bias voltage, we see that unidirectional current pulses also exist in the grid circuit. In essence, the grid behaves as a plate of a shunt-connected diode rectifier. This has two important implications in the practical operation of the Class-C amplifier. First, we can monitor the relative input drive to the amplifier by inserting a dc current meter in the grid circuit. The harder the tube is driven, the greater will be the dc component of the grid current pulses and the higher the indication on the dc grid-current meter. We realize, of course, that the presence of grid current introduces a loss not present in the Class-A amplifier, wherein input signal amplitude is not allowed to overcome the fixed negative grid bias. However, this loss is negligible compared to the great gain in plate efficiency attained by elimination of the steady dc plate current present in Class-A amplifiers. It is to be emphasized that there would be no logic in comparing the Class-C amplifier and its single-frequency resonant output tank with other types of amplifiers used for audio work. Although the audio distortion produced by Class-B amplifier tubes is readily cancelled by employing the push-pull circuit, such an expedient will not work with the Class-C amplifier. The Class-C amplifier produces virtually no proportional amplification. It performs essentially as a switching device that provides power amplification between an output and input circuit.

Grid-Leak Bias in Class-C Amplifiers

The second important manifestation of grid current is the development of automatic, or so-called grid-leak bias. This is best understood by inspecting circuitry such as that depicted in Fig. 2-7. In Fig. 2-7 we see Class-C amplifiers similar to that of Fig. 2-6 except for the important fact that the grid-bias supply is omitted. How then does our Class-C amplifier acquire sufficient negative grid voltage to protect the operation of the tube to a point beyond that corresponding to plate-current cutoff? We know that the grid-cathode circuit is supplied with a sine wave of sufficient amplitude during positive excursions to produce grid current. The positive peaks of the input voltage wave are therefore shunted down by the relatively low impedance of the grid-cathode circuit during grid conduction. This is essentially a clipping or rectifying process, and the positive excursions of the input voltage wave are thereby reduced in amplitude. On the other hand, no loading or shunting exists for the negative portion of the input voltage cycle because the negative grid does not attract electrons from the cathode. The negative grid, therefore appears as an open circuit.

Consider now the side of the grid capacitor, C, connected to the grid of the tube. Over each cycle of the input voltage wave, this side of the grid capacitor is subjected to a higher negative peak voltage than positive peak voltage. This causes capacitor C to accumulate a more or less steady charge. The charge stored in capacitor C manifests itself as a voltage difference between its plates. This capacitor voltage is polarized so that the grid is maintained negative with respect to the cathode. We observe that this dc bias is actually derived from rectification of the input signal and must,

A *Basic circuit.*

B *Modification of circuit at (A).*

C *Modification of circuit at (A).* **D** *Modification of circuit at (A).*

Fig. 2-7. Grid-leak bias in the Class-C amplifier.

accordingly, be entirely dependent upon the presence of the ac input.

Need for Discharge Path

If only capacitor C was present, this capacitor would charge up to a voltage which would bias the tube with so high a negative voltage that very little of the ac input signal would appear at the grid. Power output would consequently be very low. The resistor R, appropriately known as the *grid leak* is included as a means of providing a discharging path for the charge stored in grid capacitor, C. The rate of discharge is such that the voltage across capacitor C decreases sufficiently each cycle to enable the grid to draw current during the positive peaks of the input-voltage wave. The part-cycle grid current, in turn, is responsible for the accumulation of a dc voltage in the grid capacitor, C. Actually, for a given input-signal amplitude and value of R, an equilibrium is attained in which we have, simultaneously, some average value of negative grid bias, and some average value of grid current. Inasmuch as the grid-leak resistor, R, is inserted in the path of the external grid current, we can measure relative grid current by opening the circuit and connecting a direct-current meter in series with this resistor.

Resolving Apparent Conflict of Positive and Negative Grid

One of the difficulties in visualizing the operating principle of grid-leak bias is the apparently conflicting situation wherein we have at the same time negative grid bias greater than cutoff value, and grid current, which requires a positive grid. It should be realized that although the average value of grid voltage is a high negative value, the grid, nevertheless, is positive for a short fraction of an input cycle. During the relatively short period of grid conduction, a strong output pulse is delivered to the resonant tank in the plate circuit. This suffices for efficient excitation of strong oscillations in the tank circuit. If the ammeter and voltmeter shown connected in the grid circuit were possessed of sufficiently speedy response, we would see that the current meter would be zero most of the time, but would deflect in the positive direction for brief periodic durations. During these brief durations of high grid-current flow, the grid voltmeter would be pinned against its zero stop by the action of reverse (positive grid) polarity.

Our discussion of Class-C amplifiers has been confined to vacuum tubes. The transistor somewhat simulates the operation of a Class-C tube amplifier.

In self-excited oscillators, the transistor generally operates with characteristics somewhere between Class-B and Class-A. The bias requisite of Class-C tubes, "twice the cutoff bias" has no meaning because transistor action ceases between zero and several tenths of a volt of forward bias.

OPERATING MODES OF TRANSISTOR AMPLIFIERS

All amplifying devices are analogous in that their operating points can be established to accommodate all (Class A), one-half (Class B), or a smaller portion (Class C) of the ac signal impressed at the input. And in all instances, there are various tween modes, such as Class AB, or Class BC. Let pertains to transistors.

The circuit in Fig. 2-8(A) is that of a simple transistor amplifier. Operating bias is determine R1 is variable in order to facilitate demonstration ating modes. If R1 is adjusted so that the right amount or iui ... bias is applied to the base-emitter circuit, Class A operation will ensue, providing the amplitude of the input signal is held within limits. In Class A operation, the collector current traces out a mirror-image reproduction of the input voltage waveform. And, of course, a similar voltage-waveform will be developed across the load resistance, R_L. The operating conditions for Class B and Class C are also shown. However, the idealized graphical representation of Fig.

Fig. 2-8. Operational modes of transistor amplifiers. (A) Transistor amplifier with operating bias controllable by R1. (B) Output-current waveform as function of operating-point bias (idealized). (C) Location of operating-point bias on load line. (D) Magnified view of origin in graph (A) (Real-world situation).

s not quite true with practical transistors and more will need aid regarding Class B and Class C operation. In any event ure" Class B operation is such as to reproduce one-half cycle of the input wave.

Inasmuch as collector current, I_C, does not commence in the transistor until a forward base-emitter bias of about 0.6 volt is established, true Class B operation requires such a threshold bias. However, if R1 is made very high, or infinite, we will have what may be called Class-B operation with crossover distortion—somewhat less than a half-cycle of the input wave will be reproduced. This reasoning is more germane to audio than to oscillator work. For oscillators, this zero-bias operating condition is often referred to as Class C. In actuality, it is more on the order of Class BC. True Class-C operation of the transistor requires some cutoff bias, that is, negative V_{BE} for npn types. In practical transistor oscillators, such negative bias is not often applied. (It may, however, be developed by the rectification action of the base-emitter diode, that is by "grid-leak" action.)

More insight into these matters is provided by (C) and (D) of Fig. 2-8. In (C), it can be seen that Class-A operation implies placing the bias midway within a linear operating region. Class-B operation is seen to coincide with the bias, I_{B1}, set at +0.6 volt. (Base-current progressively increases in positive value, I_{B1}, I_{B2}, I_{B3}, etc.)

Figure 2-8(D) is a magnified view of the region around the origin in Fig. 2-8(B). This reinforces the fact that zero-bias operation of the transistor is not true Class B operation. As previously alluded, such operation in transistor oscillators is generally considered to be Class C. In practical oscillators using transistors, greatest efficiency occurs in the BC region, but greatest stability usually requires Class A operation.

MODES OF OPERATION FOR
OP AMPS AND LOGIC CIRCUITS

Somewhat different considerations prevail for op amps than for discrete devices with regard to oscillator operating-mode. An op amp is a composite device and the amplifier stages are manufactured with their own fixed-bias sources. The reason for this is that the op amp's chief mission is to perform in the linear Class-A mode. So the notion of operation in Class B or Class C is not a practical one for oscillators designed around op amps. Indeed, the op amp merits consideration when, in the interest of oscillator stability and wave purity, exceedingly linear Class-A operation is desired. Inasmuch as most op amps work at relatively low power levels, efficiency and thermal problems have low priority.

Paradoxically, the op amp can *also* function in oscillators as a switching device. As such, it behaves as a more ideal switch than discrete devices operating in their Class B or Class C modes. Such switching action in op amps is termed Class D operation. To a first approximation, it is brought about as the result of overdrive—a situation easily attainable because of the very-high gain in op amps. Indeed, steps are often taken to prevent this mode of operation. On the other hand, other op amp oscillators *deliberately* make use of the switching mode; the multivibrator is one familiar example. However, an op amp crystal oscillator may also exploit Class-D behavior, the idea being that the crystal provides the requisite frequency-stability. Such circuits are vigorous oscillators and one is not likely to encounter start-up problems, or to find that oscillation is readily killed by loading.

The reliability of such over-driven op amp oscillators appeals to many designers. Often a sine wave is not needed. But even when it is, a simple low-pass or band-pass filter or other resonant circuit can convert the square wave to a satisfactory sine wave. In a practical op amp oscillator using, say, a Hartley circuit, a distorted output wave would probably signify the over-driven mode. In order to obtain a good sine wave, it would be necessary to decrease the amount of positive feedback.

Logic devices operate like overdriven op amps, only "more so"—they are *deliberately* designed to operate in the switching mode. This is fine for RC and relaxation oscillators of the general multivibrator family. Resonant circuits and crystals are often inserted in the feedback loop to stabilize the frequency, or used as filters to produce a sine wave output. The concepts of Class-A, -B, -C operation have no relevancy here.

NEON BULB AS A SWITCHING DEVICE

The simple two-element neon bulb constitutes a novel switching device. It is but one member of a large family of gaseous diodes, most of which contain an inert gas such as neon, argon, krypton, or xenon. When low voltages are applied across the elements of such a device, the gas behaves as a fairly good insulator, permitting passage of virtually no current. As the voltage is raised, the electric force applied to the gas atoms ultimately becomes sufficient to tear outer-orbital electrons from the constraining force of the atomic nucleus. Such disrupted electrons collide with sufficient force with bound electrons to impact these from their respective orbits. The process quickly becomes a cumulative one in which an avalanche of electrons becomes available for collection at the positive element; this migration of electrons constitutes an electric current.

Additionally, atoms from which electrons have been torn have thereby acquired a net positive charge and proceed to the negative element. There they restore, temporarily at least, their lost orbital electrons. This, too, constitutes current and is additive to the total current. The total process, whereby the gas is abruptly changed from an insulator to a conductor is known as ionization. Obviously, this phenomenon can be utilized as a switching mechanism. The important feature of gaseous diodes so far as concerns studies in this chapter, is their voltage hysteresis. This denotes a difference in the voltage causing ionization and the voltage at which the diode de-ionizes. Specifically we should appreciate that the ionization voltage is higher than the deionization voltage. A gas-diode relaxation oscillator based on voltage hysteresis is shown in Fig. 2-9.

THYRATRONS

The gas diode is not a practical switching device for shock-exciting oscillations in a resonant tank circuit because there is no suitable way of synchronizing the pulsing rate with the resonant frequency of the tank. However, this device is useful in conjunction with a charging resistance and a capacitor for generating relaxation type oscillations. The origin and nature of such oscillations are somewhat different from the oscillations derived from tank circuits. A switching device with more desirable characteristics is obtained by the insertion of a control electrode for initiating ionization. Such tubes are made in several different varieties. In the simplest form, a "cold" cathode is used, that is, simply a metallic rod, not

Fig. 2-9. Gas-diode relaxation oscillator.

Fig. 2-10. Improved relaxation oscillator.

a thermionic emitter of electrons. The thyratron gas tubes are more elaborate, having, among other features, a thermionic emitter that must be supplied with filament current as is the case in vacuum tubes. An improved relaxation oscillator using a thyratron is shown in Fig. 2-10.

Some thyratron tubes contain an additional element, a screen grid that improves the operating characteristics. The chief function of the screen grid is to reduce the control-grid current, which is otherwise appreciable even before cathode-plate conduction has been established. This results in better isolation between input and output circuits. All of these controlled gas tubes suffer from the limitation that once conduction is established between cathode and plate, the control element loses further control over plate current, (Fig. 2-11). Stoppage of plate current is thereafter brought about only by lowering the plate-cathode voltage below the value required to sustain ionization. As a consequence, these tubes also find their most important use in relaxation oscillators, rather than in oscillators employing resonant tank circuits. The switching rate of all gas tubes it inherently limited by the time required for the gas atoms to deionize. This prevents generation of pulses beyond several tens of kilohertz.

THE THYRATRON INVERTER

The thyratron inverter is a relaxation oscillator of the multivibrator family. The unique features of this circuit involve the use of thyratron tubes and transformer plate loads. These circuitry fea-

A Grid voltage less than ionization value.

B Grid voltage sufficient for ionization.

C Reduction of grid voltage does not extinguish tube.

D Deionization by removing plate supply voltage.

Fig. 2-11. Control of gas tubes.

tures permit operation at much higher power levels than are readily obtained from the conventional multivibrator oscillator. Due to the time required for the gas to deionize, the frequency of the thyratron inverter is limited to several hundred hertz for mercury vapor tubes and, perhaps, several kilohertz for inert gas tubes. (Ionization time is a small fraction of deionization time and is not an important factor in the matter of frequency limitation.) A thyratron, once fired, cannot be restored to its nonconductive state by variation in grid, or starter electrode, voltage. This imposes a special problem. In order that the two tubes may alternate conduction states in multivibrator fashion, a circuit provision must be made for extinguishing the tubes by means other than the grid signal.

We recall that a conductive gas tube may be deionized by

depressing, or removing, its anode voltage. This technique is employed in the thyratron inverter. Capacitor C_x (Fig. 2-12), the so-called commutating capacitor, transfers a negative transient from the plate of the tube switching into its conductive state to the plate of the alternate tube, thereby causing deionization in the latter tube. In this way, the tube that in a multivibrator would be turned off by a grid signal is here turned off by momentary depression of plate potential below the value required to sustain ionization. On the other hand, each tube is alternately turned on in response to grid signals transferred via the cross-over feedback paths, as in the multivibrator.

The frequency of the thyratron inverter is quite sensitive to load variations. This is so because the time constant of the inductive plate circuits is affected by the effective load resistance reflected back into the primary winding of the transformer. A load with an appreciable reactive component can neutralize the effect of the commutating capacitor, thereby stopping oscillation. The output waveform tends to be an approximate square wave. A rough approach to a sine wave may sometimes be attained by inserting a choke in series with one of the power supply leads. This slows the rise and decay of the switching cycles. Contrary to the appearance of the circuit, there is virtually no filtering action due to the "tank" circuit formed by the commutating capacitor in conjunction with the primary winding of the transformer.

The resonant frequency of this LC combination is generally much higher than the switching rate of the thyratrons. Even if the inverter pulsed at the natural resonant frequency of this LC com-

Fig. 2-12. Basic thyratron inverter.

bination, the effective Q of the tank would be too low to provide much frequency selectivity. This is due to the fact that one-half of the transformer primary is always short-circuited. At the power levels generally used in electronics, thyratron inverters are not as efficient as, nor do they possess other desirable performance characteristics readily obtainable from, transistor saturable-core oscillators. However, thyratron inverters have potential application in power systems where it is desired to transform sizeable amounts of dc power to ac power. The tubes used for high power conversion resemble ignitrons, or mercury-arc tubes with starter electrodes. These tubes do not have thermionic emitters in the ordinary sense, but can, for practical purposes, be considered as big brothers to the thyratron.

SPARK-GAP OSCILLATOR

If we connect a high-voltage source of ac across a gap formed by two metal electrodes separated a small fraction of an inch, a noisy display of sparks is produced in the gap. Although the flashing display of the dielectric breakdown of the air in the gap appears to the eye as a continuous process, such is not actually the case. Rather, the arc is extinguished each time the voltage impressed across the gap passes through zero (or a relatively low value). For the ac sine wave, this occurs at double the rate of the frequency (Fig. 2-13). Thus, an arc energized by 60-Hz high voltage goes out 120 times per second. We see that such a spark gap is effectively a switch that interrupts a high voltage source 120 times/sec. If properly associated with a resonant LC circuit, these disturbances can provoke oscillations at the resonant frequency of the LC circuit (Fig. 2-14).

Fig. 2-13. The spark gap as a switching device.

74

Fig. 2-14. A spark-gap oscillator.

The LC circuit must produce many oscillation cycles for each switching pulse. Therefore, the oscillatory wavetrain is highly damped and generally decays to zero long before excitation by the forthcoming switching pulse. A considerable amount of the conversion of low-frequency to high-frequency power in the spark-gap oscillator is invested in harmonics and spurious frequencies that are not readily attenuated by practical LC tuned circuits. For this reason, the spark-gap oscillator is of historical interest only so far as concerns its use in radio communications. However, spark-gap oscillators remain useful for medical and industrial applications. For example, the spark-gap oscillator is often employed in conjunction with welding machines for initiating the arc between the welding rod and the work material. (Once the air between the welding rod and the work material is ionized by the high-voltage, high-frequency energy from the spark-gap oscillator, the low-voltage, low-frequency supply from the welder is able to establish the heavy current arc required to generate welding heat.)

NEGATIVE-RESISTANCE DEVICES

A family of important oscillation-provoking devices is represented by the so-called negative-resistance elements. Although feedback oscillators, as well as relaxation, phase shift, and spark-gap oscillators can be shown, by mathematical analysis, to be amenable to treatment with the concept of negative resistance, such oscillators are not, in the most direct and practical sense, of the negative-resistance variety. We say that an element displays negative resistance in that region of its voltage-current characteristic throughout which a decrease in voltage across the element produces

75

an increase in current through the element, or throughout which a decrease in current through the element produces an increase in voltage across the element. We see that such relationships are contrary to what we would expect from ordinary Ohm's law conduction. When such an element is operated in parallel with an LC tank circuit, it is actually possible to cancel the dissipative losses in the tank.

Sustained oscillations build up in an LC tank when it is associated with the appropriate negative resistance element. We see, that in both the mathematical and the practical sense, the term *negative resistance* is quite descriptive. Looking at the phenomenon from another angle, we can say that if ordinary, or *positive resistance* dissipates power, our negative resistance must provide power. This it does. The negative resistance element derives its power from a dc source and yields a portion of this power to the LC tank. We should appreciate that the negative resistance exists as such for the ac oscillations, not for the static dc voltage and current which define the operating point of the negative resistance element.

Concept of Dynamic Resistance

In negative-resistance devices, the negative resistance is not of the static variety such as is obtained by measuring the resistance with a dc ohmmeter. Rather, it is a dynamic parameter and exists only for ac. As such, it cannot be measured at a single operating point. The ac, for dynamic resistance, is derived by dividing a small change in voltage by the corresponding change in current. If one of these factors represents an increase and the other a decrease, the dynamic resistance is negative. In a resistor, a small increase in applied voltage would produce an increase in current; consequently, the dynamic resistance of a resistor is positive. In the dynatron tube, a negative-resistance device that is discussed in subsequent paragraphs, an increase of plate voltage from 20 volts to 25 volts might be accompanied by a plate-current change of 2.5 milliamperes to 1.5 milliamperes. This, we compute as 5 volts divided by − 1 milliampere or a negative resistance of 5000 ohms.

THE DYNATRON OSCILLATOR

The dynatron oscillator makes use of an interesting vacuum-tube characteristic for provoking oscillation in an LC tank circuit. Certain screen-grid tubes, particularly the older Types 22 and 24 tubes, exhibit a plate-current versus plate-voltage relationship such

Fig. 2-15. Negative resistance in the dynatron circuit.

as that depicted in Fig. 2-15. As plate voltage is increased from zero, we note that the plate current increases up to a certain plate-voltage value. If the plate voltage is increased further, we see that the resultant plate current decreases not merely to zero, but actually to a negative value. Over the region that this inverse current-voltage relationship occurs, the slope of the graph is opposite to those portions of the graph that depict an increase of current with a voltage increase.

The existence of the negative-resistance region results from the simple fact that throughout this region an increase in plate voltage results in a decrease in plate current. Mathematically, when plate current is caused to change from, say, a small positive value to a small negative value, the change is in the nature of a decrease. With different control-grid and screen-grid voltages, good dynatron operation can be obtained with plate currents always positive; that is, with the plate always *consuming* current from the dc plate power supply. (This does not conflict with the fact that the dynatron *delivers* ac power to the resonant LC circuit.) It should be appreciated that in ordinary Class-A amplification, the polarizing and signal potentials are such that the tube operates within the confines of the near-horizontal region of its plate-current/plate-voltage characteristic. In this way, amplification is maintained fairly linear, and the oscillation provoking tendency of the negative-resistance region does not cause trouble. But, why does this negative-resistance region exist?

Secondary Emission

In a vacuum tube, we know that thermionically emitted electrons from the cathode are attracted to other tube electrodes that are biased positive with respect to the cathode. The movement of such electrons within the tube constitutes the flow of electric current and is so indicated by a current meter inserted in a positive grid or plate lead. However, thermionic emission is not the only process whereby free electrons can be generated within a vacuum tube. In a simple vacuum-tube diode, we know that increased plate voltage results in increased plate current, (assuming operation is below plate-current saturation). The increased plate voltage not only attracts more electrons per unit time from the cathode area, but imparts greater acceleration to these electrons. If the plate voltage is high enough, the attracted electrons have sufficient kinetic energy to impact orbital electrons from the surface atoms of the plate metal. Electrons thus freed from their atomic bonds are called secondary electrons to distinguish them from the primary electron emission from the heated cathode. (No appreciable elevation of plate temperature is required for secondary emission.) Significantly, a single primary electron can liberate two or more secondary electrons. In the diode, or Class-A triode, secondary emission from the surface of the plate is not of any great consequence because such electrons are simply attracted back to the positive plate. Let us investigate what can happen in certain tetrodes.

Reason for Dynatron Property of Negative Resistance

Suppose that considerable secondary emission is being produced at the surface of the plate in a tetrode or screen-grid tube, and that the screen-grid voltage is higher than the plate voltage. We see that it is only natural that the secondary electrons accelerate to the screen grid rather than return to the plate. This being the case, screen-grid current must increase at the expense of plate current. (Secondary electrons which leave the plate constitute a flow of current away from the plate.) The higher the plate voltage, to an extent, the more pronounced is the generation of secondary electrons. We have the condition wherein increased plate voltage produces a decrease in plate current. This condition manifests itself as negative resistance when the tube is properly associated with an LC tank circuit. At low plate voltages, insufficient secondary emission occurs to produce the negative-resistance characteristic. At plate voltages approaching and exceeding the value of the positive screen-grid voltage, the secondary electrons return in every greater numbers to the plate. These two conditions establish the limits of the negative-resistance region.

We see that only a small portion of the plate-current versus plate-voltage relationship is suited for dynatron operation. Unfortunately, modern versions of the Types 22 and 24 tubes are manufactured under a process designed to inhibit generation of secondary electrons. Pentode type tubes have a so-called suppressor grid located between the screen grid and the plate. This added element is operated at cathode potential. Electrons liberated from the plate bear a negative charge as do the primary electrons emitted from the cathode. Therefore, the negative suppressor grid repels the secondary electrons, causing them to return to the plate, rather than accelerate to the screen grid. This results in the pentode tube having a more extensive region over which linear amplification can be obtained, but it is obvious that in the pentode we have eliminated the very mechanism required for dynatron negative resistance.

TRANSITRON OSCILLATOR

Paradoxically, although the pentode will not perform as a dynatron oscillator, it has more or less replaced the tetrode dynatron tube as a negative-resistance device for provoking oscillation in an LC tank circuit. One of the disadvantages of the dynatron is the dependency upon secondary emission. This varies from tube to tube and with aging. Indeed, it is quite difficult to obtain satisfactory dynatron action with present day tetrodes, for these are processed during manufacture to make secondary emission less copious than in tubes of earlier production. The pentode can be connected as a transitron oscillator to provide a negative-resistance region which, insofar as concerns the resonant tank, behaves very similarly to the dynatron. The essential difference between the two circuits is that negative resistance in the dynatron involves the relationship between plate voltage and plate current, whereas negative resistance in the transitron involves the relationship between screen-grid voltage and screen-grid current.

Reason for Transitron Property of Negative Resistance

In Fig. 2-16(A) we see that the suppressor grid of a pentode connected to exhibit transitron characteristics is biased negative with respect to the cathode. However, the actual value of the negative suppressor voltage is determined by the screen-grid voltage. This is so simply because the screen-grid voltage supply and the suppressor bias supply are connected in series. If we increase the screen-grid voltage, the suppressor grid bias becomes relatively less negative. This being the case, electrons passing through the screen-grid apertures are less likely to be repelled back to the screen grid; rather, such electrons are more likely to reach the plate where they add to plate current, but at the expense of screen-grid current. The converse situation likewise operates to establish the negative-resistance region. That is, if screen-grid voltage is decreased, suppressor-grid bias becomes relatively more negative, and electrons passing through the screen grid are deflected back to the screen grid, thereby adding to screen-grid current at the expense of plate current.

Thus, we see that screen-grid current changes are opposite in direction to screen-grid voltage changes. Such a negative resistance does not exist over the entire range of screen-grid current versus screen-grid voltage characteristics, but only over a small portion of the total characteristics. We see this must be so because suppressor voltage cannot become positive, for then it would draw current as the result of attracting electrons that otherwise would be collected by the plate. At the other extreme, suppressor voltage cannot become too negative, for then it would cut off plate current by deflecting electrons back to the screen grid. For the transitron action to exist, the plate must be allowed to participate in apportioning the total space current. This it can do only when suppressor voltage is neither too low nor too high.

Function of Capacitor in Transitron Oscillator

The capacitor, C_2, in the transitron oscillator circuit of Fig. 2-16(B), communicates the instantaneous voltage developed across the tank circuit to the suppressor, where it is superimposed upon the fixed negative bias voltage. It is educative to observe that the suppressor grid exerts stronger control within the negative-

Fig. 2-16. Negative resistance in the transitron circuit.

resistance region than does the screen grid. Thus, when the voltage impressed at these two elements increases, we expect screen grid current to increase, for such would be the case at an anode, which the screen grid simulates. However, the simultaneous decrease in negative voltage applied to the suppressor has a more pronounced effect upon the screen-grid current. As explained, the additional electrons that are now allowed to pass to the plate would in ordinary circuits have been collected by the screen grid. Raising the voltage at the screen grid in a transitron circuit causes a decrease in screen-grid current. Like the dynatron, the transitron is a two-terminal oscillator, it being only necessary to connect the two terminals of a parallel-tuned LC circuit. Some of the tubes that make satisfactory transitron oscillators are the 6AU6, 6BA6, 6J7, and 6K7.

Oscillation occurs when the dynamic screen-grid resistance is numerically equal, but of opposite sign to the equivalent resistance of the resonant LC tank; that is:

$$r_{SG} = - \frac{L}{(R_S)\,(C)}$$

R_S, as we recall, is the total series resistance of L and C.

THE UNIJUNCTION TRANSISTOR

The unijunction transistor is a semiconductor switching device with characteristics somewhat suggestive of thyratron action. The physical configuration of the unijunction transistor is shown in Fig. 2-17(A). It is essentially a pn structure in which two base connections, rather than the single base connection of ordinary rectifying junctions are provided. We know that reversal of applied polarity changes conduction from a high to low value in the ordinary junction diode. In the unijunction transistor, which is essentially a double-base junction diode, reversal of polarity between the centrally located p, or emitter element, and the n-type bar occurs abruptly, giving rise to a switching process in the external circuit. Consider the situation depicted in Fig. 2-17(B). When the switch is initially closed, the p element will assume a potential relative to base No. 1, B1, equal to approximately half of the voltage applied by the battery connected to the two bases. This is true because the resistivity of the semiconductor bar is uniform along its length. At first, the major portion of the p element will be substantially in the nonconducting state because of insufficient voltage gradient between it and the n material of the semiconductor bar. However, the lower tip of the p element will be slightly in its forward-conduction region, that is, the lower tip will act as an emitter of positive charges (holes). (This is because the lower tip is off-center.)

The initial conductive state following closure of switch S is not stable. The positive charges injected from the lower tip of the p element rapidly "poison" the lower half of the semiconductor bar. These charges are collected by negatively polarized base No. 1. The movement of these charges through the lower half of the bar constitutes current. In essence, the conductivity of the lower half of the bar is increased. This, in turn, shifts the distribution of potential drop within the bar so that the entire p element finds itself rela-

Fig. 2-17. Conduction states in a unijunction transistor.

tively positive with respect to adjoining n material of the bar. As a consequence, the entire p element becomes an injector of positive charges, thereby increasing conductivity of the lower half of the bar to a high degree. This occurs with extreme rapidity.

This is the stable end result of the circuit as shown in Fig. 2-17(D). The p element, that is, the emitter, in conjunction with the n material of the lower half of the bar and the ohmic base connection (No. 1), behaves as an ordinary pn junction diode passing heavy current under the operation condition of forward conduction.

Instead of the battery, V/2, let us substitute a capacitor and a charging resistor (see Fig. 2-18). For a time, the capacitor will charge much as it would if the p element and base No. 1 acted as a very high resistance. When the capacitor voltage attains the value

Fig. 2-18. Basic unijunction relaxation oscillator.

F (approximately one half the base-to-base voltage), depicted in the curve of Fig. 2-19, the cumulative process of increased charge injection and spread of charge-emitting surface will be initiated. The capacitor will then "see" a relatively low resistance and its stored voltage will be rapidly depleted. During this discharge, the avalanching of charge carriers continues within the lower half of the bar until the capacitor voltage is quite low—in the vicinity of two volts. In other words, insofar as concerns the external circuit connected between the p element and base No. 1, there is a voltage-hysteresis effect analogous to that of the neon-bulb relaxation oscillator wherein ionizing voltage is higher than extinction voltage. In any event, when the capacitor discharges to a sufficiently low voltage, injection of positive charges into the bar ceases, and the lower half of the bar reverts to its initial state of high resistivity. The capacitor charge-discharge cycle then repeats. In some cases the capacitor-charging resistor may be dispensed with. The capacitor then receives its charging current through the reverse-conducting upper half of the bar, which acts as a constant-current generator. This causes the capacitor to charge linearly with respect to time and the resultant sawtooth has a suitable shape for many sweep circuit applications. This mode of operation imposes the practical limitation that the pulse repetition rate cannot be made as high as when a charging resistor is used.

TRIODE INPUT AS NEGATIVE RESISTANCE

There is an important circuitry family of triode-tube oscillators, the operation of which is generally ascribed as being due to feedback via the internal plate-grid capacitance of the tube. Actually, such oscillators are more closely related to those of the negative-resistance family than to the feedback group. This is so despite the fact that it can be shown mathematically that even "true" feedback oscillators present negative resistance to their resonant tanks. In those oscillators we classify as of the feedback variety, the negative-resistance property is intimately associated with the presence of the resonant tank. However, in the so-called negative-resistance oscillator, the negative-resistance property is displayed by the tube or other device with or without the resonant tank.

There considerations bring us to the interesting fact that the grid-cathode circuit of a "grounded-cathode" triode amplifier can be made to appear as a negative resistance. This is brought about by making the plate load inductive (see Fig. 2-20). For a certain range of plate-circuit inductance, the grid circuit is capable of supplying power to a resonant tank. The action involved here is not easy to visualize without considerable mathematical analysis. Nevertheless, the effect is an important one and is a basic principle in the operation of the tuned-plate tuned-grid oscillator, cer-

Fig. 2-19. Current-voltage relation in Base 1-to-emitter diode.

Fig. 2-20. Under certain conditions the input circuit of a triode may appear as a negative resistance.

tain variations of the Hartley oscillator, and the Miller crystal oscillator. These oscillators will be considered in detail in our discussion of practical devices. Although useful negative input resistance does not appear for just any value of plate circuit inductance, it is found in practice that a broad range of inductance values suffice to bring about useful values of negative input resistance. This is particularly true when the Q of the inductive plate load is high. Throughout the range of inductance over which the input circuit contains negative resistance, the effect of varying the inductance is to change the numerical value of the negative input resistance. The value of the input resistance can be found with the following formula:

$$R_{IN} = - \frac{1}{2\pi f C_{gp} K \sin \theta}$$

where

R_{IN} is the input resistance in ohms,
C_{gp} is the effective grid-plate capacitance, expressed in farads,
f is the frequency in hertz,
θ is the phase angle of the plate load,
K is the voltage gain of the tube.
Note: θ and $\sin \theta$ are positive for an inductive load. This makes R_{IN} negative.

(The input capacitance of the tube also changes with variations in the nature and magnitude of the plate load. This is not relevant to our particular discussion, however.)

86

THE SATURABLE MAGNETIC CORE

The saturable magnetic core is at once an oscillating element and an oscillation provoking device. Nevertheless, it requires the association of a controllable switch such as a transistor in order to produce oscillation. The generated waveform is nonsinusoidal; such an oscillating system is of the relaxation type.

The best core material for this type of oscillation exhibits a rectangular hysteresis loop, such as that shown in Fig. 2-21(B). Suppose the battery circuit has just been completed. Conduction will start in the emitter-collector diode of one or the other transistors, but not in both because whichever one first conducts induces a reverse bias in the base-emitter diode of the opposite transistor, which prevents conduction. Due to inevitable circuit and transistor imbalances, it is highly probable that such a starting mechanism will follow closure of the battery circuit switch. Sometimes, in order to ensure starting reliability, one transistor is provided with a small amount of forward bias, or each is provided with different forward biases. Forward bias is readily provided by inserting a small resistance in series with the base of a transistor, and a second, higher resistance between collector and base terminals.

OSCILLATION IN THE SATURABLE-CORE CIRCUIT

In any event, we will assume that conduction has commenced in one of the transistors, say the one labeled Q1 in Fig. 2-21. This conduction results in a rapidly increasing current through the portion of the winding between CT and the emitter of Q1. As this current increases, it induces *two* important voltages. One of these appears across the portion of the winding between emitter and base of Q1. This voltage is of such polarity as to drive Q1 into increasingly heavy conduction. The other voltage appears across the portion of the winding between emitter and base of Q2 and is of such polarity as to clamp Q2 in its nonconducting state. The rapidly increasing current through the portion of the winding which feeds the emitter of transistor Q1 ultimately causes abrupt magnetic saturation of the core. When this occurs, electromagnetic induction of the voltages controlling the bases of the two transistors ceases. However, these voltages cannot instantaneously return to zero due to the energy stored in the magnetization of the core. Accordingly, the conduction states of the two transistors remain for a time as they were immediately preceding the onset of core saturation. However, once the collapse of the magnetic field gets well under way,

Fig. 2-21. The saturable core as an oscillation-provoking device.

sufficient reverse-polarity voltages are electromagnetically induced in the base-emitter portions of the winding. This reverses the conduction states of the transistors, turning Q1 off and Q2 on. The conduction of the emitter-collector diode of Q2 now repeats the cycle of events, but with core saturation ultimately being produced in the opposite magnetic sense. After an interval following this second saturation process, the transistors again exchange conduction roles, and the core is on the way to saturation in the same magnetic sense as originally considered. Thus, the transistors alternate states: when Q1 is "on," Q2 is "off" and vice versa. Two additional saturable core oscillators are shown in Fig. 2-22. Saturation flux density, B_s, for various core materials is given in Table 2-1. The following design data applies to Fig. 2-22 and to Fig. 2-21 as well:

$$N = \frac{E \times 10^8}{4f\,B_s\,A}$$

Fig. 2-22. Saturable-core oscillator circuits.

where,

N is the number of turns from CT,

E is the battery voltage,

f is the frequency of oscillation in hertz,

B_s is the flux density at saturation in lines per sq cm,

A is the cross-sectional area of core in sq cm.

The number of turns in the windings indicated by "n" in Figs. 2-22(A) and (B) generally is equal to N/8 to N/3.

The frequency of oscillation is determined by the battery voltage, the number of turns in the portion of the winding that feeds the emitters of the transistors, the flux density at which saturation occurs, and the cross-sectional area of the core. However, once the

Table 2-1. Saturation Flux Densities, B_s, for Various Core Materials*.

Core Material	Saturation Flux Density in Kilogauss (Thousands of lines/sq. cm)
60-Hz Power Transformer Steel	16–20
Hipersil, Siletcron, Corosil, Tranco	19.6
Deltamax, Orthonol, Permenorm	15.5
Permalloy	13.7
Mollypermalloy	8.7
Mumetal	6.6

*When using these B_s values, make certain that the core area, A, is expressed in square centimeters.

oscillator has been constructed, battery voltage is the only frequency-governing parameter that requires consideration, the others being fixed by the number of turns placed on the core, and the nature and geometry of the core material. This type of oscillator is very efficient, efficiencies exceeding ninety percent being readily obtainable. We note that there is no need for high power dissipation in the transistors. When a transistor is "off" its collector voltage is high, but its collector current is virtually zero. Conversely, when a transistor is fully "on" its collector current is high, but its collector voltage is very low. In either conduction state, the product of collector voltage and collector current yields low wattage dissipation in the collector-emitter diode. We see that one of the advantages of rectangular hysteresis loop material is that a core of such material permits rapid transition between "off" and "on" states. This is desirable, because when a transistor is neither fully on nor fully off, the product of collector voltage and current can be high. If the transition is not rapid, efficiency will be lowered and the transistors will be in danger of being destroyed by generation of excessive internal heat. Silicon steel such as is used in 60-Hz transformers is often employed to save costs. However, less power output and lower efficiency are the result.

We often see these saturating-core oscillator circuits with a capacitor connected across the load winding, or even across the primary winding. However, no LC tank is formed by the addition of such capacitors and the frequency is not changed thereby. The capacitor is added in order to absorb voltage transients that can destroy the transistors. These transients, or "spikes," are due to the abrupt saturation of the core. The frequency of these oscillators is generally limited, at least in units designed for high power, by the frequency capability of the power transistors. However, frequencies on the order of several hundred hertz to around 100 kHz are readily attainable and are well suited for high efficiency and small, light-weight cores.

THE ELECTRON BEAM IN A VACUUM

In most of the electrical oscillating elements and oscillation provoking devices we have considered, the described action resulted from the controlled or guided flow of electrons in metallic conductors of various geometric configurations. In waveguides and resonant cavities, we found that the electron was somewhat freed from the restrictions ordinarily prevailing in low-frequency "ohmic" circuits. In these microwave elements, much of the energy associated with the electron manifested itself as electric and magnetic fields external to metallic conducting surfaces. Thus, the electron movement, that is, the electric current in the walls of the waveguide, is at once influenced if we insert an obstruction within the aperture of the guide. Such an obstruction disturbs the patterns of the electric and magnetic force fields. We can go a step further than this. We can free the electron from confinement in matter completely, in which case it becomes highly susceptible to manipulation by various force fields. This is most directly accomplished by some arrangement involving a thermionic cathode and a positively biased anode, the basic objective being to create a beam of electrons in a vacuum.

Historically, the X-ray tube has considerable relevance. In this tube, the electrons are imparted tremendous acceleration by the extremely high potential difference between anode and cathode. The energy thereby acquired by the electrons is sufficient to produce dislocations in the orbital electrons of the atoms in the anode metal. Inasmuch as mass and compliance are associated with these orbital electrons, a wave motion in the form of an oscillating electromagnetic field is propagated into space. Although the frequencies of these waves are far beyond those generated by the oscillators dealt with in this book, we should realize that the X-ray tube utilizes the basic principle of all modern microwave tubes. This principle is the conversion of the energy in an electron beam to oscillatory power at a desired frequency.

THE MAGNETRON

The magnetron, like the X-ray tube, is a vacuum diode in which thermionically freed electrons are imparted high kinetic energy by an intense electric accelerating field. In the X-ray tube, the staccatolike rain of these accelerated electrons bombard the anode at appropriate intervals to sustain electron oscillation between orbits of the anode atoms. In the magnetron, the objective is to excite oscillations in microwave cavities. We cannot do this by the direct action of electrons simply accelerated by high voltage. Rather, we require the passing motions of high energy electrons at the vicinity of a cavity. This must occur at intervals at, or very near to, the natural resonant frequency of the cavity. The electric field of the electron can, under such circumstances, interact with the oscillator electric field of the cavity in such a way that a synchronized excitation is applied to the cavity once per cycle. In the magnetron, this is achieved by causing the electrons in the cathode-anode gap to describe orbital motions. Let us see how this is done.

Motor Action on Electrons

We know from electrical theory, that a so-called "motor" action occurs when a current-carrying wire is exposed to a magnetic field. This, indeed is the principle underlying the operation of electric motors. It is well to appreciate that in the electric motor, the repulsive force is not actually exerted upon the physical wires imbedded in the armature slots, but rather upon the moving electrons within the wire. This being true, we are not surprised that the same motor action may be imparted to moving electrons in a vacuum. The magnetron employs a powerful magnet for this very purpose (Fig. 2-23A). The field is oriented perpendicular to the plane of electron emission from the cathode. At low plate voltages, the motor action results in small tight orbits and the electrons do not reach the plate at all. This is attested by the negligible plate current which exists at low plate voltages (see Fig. 2-23 B). At higher plate voltages, the magnetic field has less deflective power and the electron orbits are consequently larger.

In fact, some of the electrons simply "fall" into the plate metal and are collected as in the ordinary vacuum diode. The collection of such electrons constitutes plate current and is the source of use-

A *Diagram of basic construction.* **B** *Plate-current/plate-voltage curve.*

Fig. 2-23. Multicavity magnetron.

Fig. 2-24. Basis for oscillation in the reflex klystron.

less plate dissipation which lowers the efficiency of the tube. Other electrons, however, orbit past cavity apertures at just the right time to excite and sustain oscillations in the cavities. These electrons, together with others which make relatively small orbits by virtue of low initial velocities of emission, return to the cathode where their remaining energy is dissipated as heat. Indeed, in high power magnetrons, it is sometimes possible to turn off the heater current once the magnetron has become operative, the energy of such returned electrons being sufficient to maintain the cathode at the temperature required for thermionic emission. It is quite common to reduce the heater current once the magnetron is oscillating. We see that magnetron action is dependent upon synchronization of the orbital cycles of the electrons with the resonant frequency of the cavities. Oscillation is excited and sustained by interaction of the electric field in the oscillating cavity and the electric field of the passing electron. It is not necessary or desirable that the electrons be collected by the anode structure. The microwave energy is extracted by a loop inserted in one of the cavities. This couples the total oscillatory energy of all cavities to an external load because all cavities are "electronically coupled" by the modulating action of the orbiting electrons.

THE REFLEX KLYSTRON

The reflex klystron, shown in Fig. 2-24, employs a somewhat different stratagem to extract energy from an electron beam in the form of microwave oscillation. The anode of the klystron is a resonant cavity that contains perforated grids to permit accelerated electrons to pass through and continue their journey. Such electrons are not, however, subsequently collected by a positive electrode. Rather, they are deflected by a negatively polarized "reflector" and are thereby caused to fall back into the cavity grids. The operational objective of the tube is to have such electrons return to the cavity grids at just the right time to reinforce the electric oscillatory field appearing across these grids. When this situation exists, oscillations are excited and sustained in the cavity. Microwave power is coupled out of the cavity by means of a loop if coaxial cable is used, or simply through an appropriate aperture if a waveguide is used for delivering the power to the load. After the kinetic energy of the electrons has been given up to the oscillatory field of the cavity, the spent electrons fall back to the positive-biased control grid where they are collected, thereby adding to control grid current. If the tube is not oscillating, a relatively high number of electrons are deflected by the retarding field of the reflector with sufficient energy to pass through the cavity grids, thence to be collected by the control grid. However, when oscillations are sustained in the cavity, the falling electrons yield most of their energy to the oscillating electric field appearing across the cavity grids. Such electrons are subsequently collected by the cavity grids, which in this function behave as the plate of an ordinary diode. Inasmuch as the spent electrons do not fall into the positive field of the control gird, a profound dip in control-grid current accompanies the onset of oscillation within the cavity.

Bunching of Electrons Returned to Cavity Grids

Depending upon their time of arrival from the cathode, different electrons will be imparted different accelerations if oscillations exist in the cavity. Some electrons will experience violent attractions when the oscillating electric field of the lower cavity grid is at its positive crest. Others will be subjected to less intense attractive force because they approach the lower cavity grid at a time of the oscillatory cycle when it is not so highly positive. (During oscillation, the dc voltage level of the cavity grid is effectively modulated at the oscillatory rate.) As a consequence, some electrons will be projected higher into the retarding field of the reflector than others. Nevertheless, electrons projected upwards through the cavity grids can be caused to return to the cavity in groups, or "bunches" by adjustment of the reflector voltage. In order for electrons with different penetrating distances into the retarding field to simultaneously return to the cavity grids, such electrons must obviously possess different velocities. The electrons in the region between the reflector and the upper cavity grid are said to be *velocity modulated*. When, under influence of proper reflector voltage, bunches of such electrons return to the cavity grids, the energy they acquired during their fall from the retarding field is extracted by the oscillating electric field across the cavity grids. We see a cumulative process: oscillation produces velocity modulation, and the bunching due to velocity modulation delivers synchronized energy to the cavity grids, thereby sustaining oscillation. In the reflex klystron, as in the magnetron, microwave oscillation is provoked in cavities by the influence of timed electrons. In both instances the oscillatory power is derived from the energy of electrons accelerated in a vacuum.

TRAVELING-WAVE TUBES AND THE BACKWARD-WAVE OSCILLATOR

A third method of extracting energy from an electron beam to produce microwave oscillatory power is utilized in the backward-wave oscillator. This device is a specialized member of a family of microwave tubes known as traveling-wave tubes. We will discuss amplification in the traveling-wave tube because this is the basic process that must be understood in the operation of the backward-wave oscillator. In the traveling-wave tube, a beam of

Fig. 2-25. Traveling-wave tube.

electrons is projected through the center of a long helix (see Fig. 2-25). It is desirable that the electrons should pass close to, but should not be collected by, the positively biased helix structure. This requires formation of an electron beam of very small diameter. This is not, in itself, sufficient, for an electron beam tends to diverge due to mutual repulsion of the electrons comprising the beam. In order to constrain these electrons within the inner diameter of the helix, a strong longitudinal magnetic field is applied parallel to the axis of the helix. The motor-action experienced by the divergent electrons interacting with this field imparts forces urging them toward the geometric axis of the helix. Such focusing of the electron beam is the first requisite of all traveling-wave tubes. Optimum beam focusing is accomplished by the alignment of the magnet structure, which results in minimum helix current. Let us, now, consider the function of the helix.

Function of Helix as "Slow Wave" Structure

The helix behaves essentially as an "open" wave guide. Although not commonly employed in microwave or vhf practice, a single conductor can guide the progress of electromagnetic waves. The single wire lead-in for television antennas (the so-called G-string transmission line) is a wave guide of this nature. Despite the fact that a signal may propagate along the surface of the helix conductor at approximately the speed of light, the progress of the waves along the axis of the helix will be only a fraction of this speed, being governed by the winding pitch of the helix. Of prime importance, the speed of the electrons within the beam can be made nearly the same as the speed of the signal waves traveling axially down the length of the helix. When this condition prevails, we have an interaction between the electron beam and the signal. This interaction is such that the signal increases its power level during its axial progress from the cathode end to the collector end of the helix. This power is extracted from the electron beam. At the output end of the tube, the electrons are collected by a positively polarized electrode and the amplified signal is coupled from the end of the helix to a coaxial fitting, or to a waveguide aperture. In such a traveling-wave amplifier, it is important that the amplified signal is not reflected back to the input end of the helix in such a way as to promote the buildup of oscillation. This is prevented by making at least as portion of the helix "lossy." Reflected waves are then attentuated. Of course, forward-traveling waves suffer attenuation too, but it will be shown that these waves grow in amplitude by absorbing energy from the electron beam.

The traveling-wave amplifier makes an excellent oscillator if a microwave cavity is externally inserted in a feedback loop between output and input. The cavity then determines the frequency of oscillation. The traveling-wave amplifier is relatively insensitive to frequency, this being its much heralded feature. This is because the helix behaves much like a transmission line terminated in its own characteristic impedance, i.e., it is nonresonant. Traveling-wave amplifiers commonly cover frequency ranges in the vicinity of two to one. Often the limiting factor to the extent of frequency response is the input and output coupling mediums. (There is considerable practical difficulty encountered in preventing coupling helixes, loops, or probes from becoming reactive when operation over extremely wide frequency ranges is attempted.) Of course, we wish to know what makes the axial component of the helix wave grow as it journeys adjacent to the electron beam down the tube.

Bunching of Electrons in Beam

Referring to Fig. 2-26, we see that the signal wave in its propagation along the helix reacts upon the electron beam. Specifically, the beam is alternately speeded up and slowed down; we say it is "bunched." When the electric field of the signal wave increases in the positive direction, the electrons in the beam are speeded up. When the electric field of the signal wave increases in the negative direction, the electrons in the beam are slowed down. When the beam electrons are speeded up, a region of high electron density is formed; when the beam electrons are slowed down, a sparse region of low electron density is formed. The average speed of elec-

Fig. 2-26. Wave generation in a traveling-wave tube.

trons in traversing from the gun end to the collector end of the tube is not changed by this bunching process. Such electrons are speeded up to the same extent they are slowed down. The physics of kinetic energy has something very definite to say regarding such a situation; no net energy is imparted to, or extracted from, the electron beam as the result of such bunching. However, the dense electron regions have stronger magnetic fields than the sparse electron regions. This means that the helix is subjected to a change in flux, which from basic electrical principles, we know must induce a voltage wave in the helix. This induced or "reinduced" wave grows in power level during its progress along the helix. How do we account for this action?

Progressive Increase in Bunching Density

We note in Fig. 2-26 that, commencing from the gun end of the helix, each successive region of bunched electrons is denser than that preceding it. This is so because, once bunched, electrons tend to remain so during their transit down the tube. Consequently, a given bunched region is repeatedly subjected to the bunching process as it interacts with the periodic electric field of the signal wave. The bunching thereby becomes cumulative, with electron density increasing as the electrons in the beam are bunched during their transit from the gun to the collector end of the tube. In similar fashion, the sparse regions become less dense with successive interactions with the electric field of the signal wave. The natural consequence of this situation is that the electromagnetic induction of the second helix wave, the one we previously referred to as being "reinduced," becomes more pronounced as we approach the collector end of the helix. In this way, the power level of the second helix wave grows from one induction cycle to another. This growing wave does extract energy from the electron beam. In giving up energy to the helix wave, the beam electrons are slowed down. In order to effect an energy transfer, it is necessary that the reinduced wave be propagated slightly slower than the average speed of beam electrons. This condition is readily brought about by adjustment of the dc helix voltage. (It so happens that when the signal wave and the electrons have identical propagation speeds, the reinduced wave, i.e., the growing wave, lags behind the average electron transit time.) Although basically sound, our analysis has been simplified in that we considered but *one* growing wave. This growing wave itself velocity modulates the beam to produce electron bunches, which in turn reinduces another growing wave on the helix. This process occurs numerous times so that the ultimate growing wave is a composite of many such waves.

Backward-Wave Oscillator

In the backward-wave oscillator, we permit reflection of the composite growing wave back to the input end of the helix. Oscillation occurs at the frequency at which the phase of returned signals most strongly reinforces the helix wave at the input end of the tube. No external cavity or other resonant device is necessary to tune the oscillations. Rather, the frequency of oscillation is governed by helix voltage. We see that traveling-wave tubes, in common with klystrons and magnetrons, effectively *use* the transit time of electrons. Conversely, in ordinary vacuum tubes, transit time of the electrons in the cathode-plate gap establishes the upper limit to high-frequency performance.

OSCILLATOR THEORY IN TERMS OF THE UNIVERSAL AMPLIFIER

The largest class of oscillators make use of an active device capable of providing power amplification, a resonant circuit, and a feedback network or provision. The unique association of these functional elements identifies the oscillator and generally is responsible for its operating characteristics. Inasmuch as there are many oscillator circuits, as well as a number of active devices, it would greatly simplify our study of oscillators if it were not necessary to deal individually with the large number of permutations acruing from these facts. It fortunately turns out that most oscillators can be made to work with most active devices. To a considerable extent, oscillators "don't care" what kind of active device is employed.

This is not to say that optimum performance will necessarily attain from a blind selection of the active device, or that one can always directly substitute one active device for another. However, with a little experimentation, or calculated modification of component values, most oscillators are capable of satisfactory performance with two, three, or more active devices, any of which may have been chosen for a variety of reasons. This makes it unnecessary to treat the theory of oscillation separately for different active devices and leads to an elegant conceptual approach to the subject.

Dealing with a "universal" active device such as the amplifier enables us to focus attention on the circuitry aspects of oscillators. Once the basic principles of an oscillator circuit are grasped, the designer or experimenter is then free to consider an npn or pnp transistor, an N-channel or P-channel JFET or MOSFET, various type of op amps, or an electron tube. The choice may be governed or influenced by considerations of frequency, power level, availability, cost, or system compatibility. But, a single basic theory will apply, no matter what choice is made. Moreover, it should be noted that most active devices are basically three-terminal devices. Even those with additional terminals still operate as amplifiers via the interaction of three "main" terminals or electrodes. Dual-gate MOSFETs and screen-grid tubes are examples of such cases.

Figure 2-27 shows the symbol of a universal amplifier this book will use in order to exploit the aforementioned facts. The triangle symbol has long been used in block diagrams to depict various types of amplifiers. Although it is also well known as the symbol of a particular active device—the op amp—its proper use in this connection requires the input terminals to be labeled + and –, or noninvert and invert. The technical literature generally supports the use of the unmarked triangle as a representative amplifier. In other words, the triangle symbolizes the function of amplification. It is in this sense that it will be used to facilitate the explanation of various oscillator circuits. Later, specific symbols of various active devices will be used in practical oscillator circuits to serve as examples that serve as guides to actual applications.

A number of active devices that can be represented by the universal active device or universal amplifier are shown in Fig. 2-28.

Fig. 2-27. The universal amplifier symbol. Inasmuch as most oscillators can be made to operate with a *variety* of active devices, it is too restrictive to discuss the operating principles in terms of a single device. The triangle symbol is commonly used to depict all kinds of amplifiers and is, therefore, appropriate for use with oscillator circuits.

Fig. 2-28. Active devices that can be used in oscillator circuits. Those shown on the right are readily usable in common oscillators such as the Colpitts circuit of Fig. 2-29. Logic gates are often used in various oscillator circuits also.

These probably embrace the vast majority of devices presently used in oscillators. One could, however, postulate others. For example, thyristors such as the SCR and the PUT could conceivably be included. Also there are new solid-state devices being developed that could assume importance in oscillator applications. An example is a hybrid device in which a power MOSFET and a power bipolar transistor are monolithically joined so that the net device has a field-effect control terminal and a pn junction output section. Also, it is possible to cite other logic devices in addition to those illustrated that can be used in oscillator circuits. The Schmitt trigger, for example, is sometimes seen in crystal oscillator circuits used for clocking logic systems.

SOME CONSIDERATIONS IN THE SELECTION
OF SEMICONDUCTOR DEVICES FOR OSCILLATORS

Figure 2-29 shows a Colpitts oscillator configured about an active device. The active device can be any of the symbolic representations shown. This is because they all produce power-gainand are, essentially, three-terminal devices. (Tetrode and pentode tubes, as well as dual-gate MOSFETs can be viewed as basically three-terminal amplifying devices with auxiliary control-elements.) However, there are significant differences in performance, convenience, and cost among the active devices. Some excell at high frequencies, while others would merit consideration if we were striving for high power-levels. With some, it is easier to stabilize frequency against the effects of temperature, operating voltage, or aging. Some are more immune to physical and electrical abuse. The impedance-levels of some are less degrading to resonant circuits than are others. And, naturally, some will be found more compatible to a particular system or application than others.

Some examples will illustrate factors likely to influence the choice of an active device for an oscillator. To begin with, if it is desired to attain rapid warm-up and to conserve space and weight, the electron-tube would necessarily have to give way to solid-state elements. If high input-impedance is required, the bipolar transistor would generally not be as good a choice as the MOSFET or JFET. And, in this instance, power MOSFET are commonly available, whereas power JFETs remain the objects of overseas experimentation. Linear op amps can be configured into very precise oscillators because their high gains enables the simultaneous use of positive and negative feedback. Thus, an op amp would be a good device to use in a sine wave oscillator where the emphasis is on

Fig. 2-29. A basic Colpitts oscillator shown with a three-terminal amplifying device. The implication of such a schematic diagram is that a wide variety of active devices can be employed in the circuit. The triangle is, accordingly, the symbol of the so-called "universal amplifier."

wave purity rather than power output.

Sometimes the selection of the active element is based upon more subtle considerations than, say, frequency or power capability. For example, the design of a solid-state dip meter (intended to perform similarly to the tube-type grid-dip meter) might not work satisfactorily with a bipolar transistor because the bias current of the transistor tends to mask the rectified oscillator current. And a MOSFET lacks an input diode needed to provide the "grid" current. It so happens, however, that a JFET has an input diode somewhat analogous to the grid-cathode diode of the tube. The analogy is close enough so that the JFET can almost directly substitute for tube circuits in this application. But for most other oscillator purposes, it is undesirable to operate so that the input diode is driven into forward conduction.

The nearly universal use of bipolar transistors for all kinds of oscillators may not always reflect the best wisdom of choice. For example, a bipolar transistor oscillating at a frequency well down on its current-gain characteristic may cause trouble with low-frequency "parasitics" because of its relatively high gain at low frequencies.

The JFET is a very popular device for almost direct substitution in the crystal oscillator circuits that were long used with tubes. Many crystals are intended for oscillation in the parallel-resonant mode and are stipulated for a certain frequency with the proviso that a capacitance of 32 pf be presented by the oscillator circuit. Because of the relatively-high input impedance of the JFET, this condition is readily met. Moreover, the input impedance of the JFET is less influenced by temperature or by dc operating-voltage than is the case with bipolar junction transistors. Thus, the frequency stability of the crystal tends to be good with very little effort expended in circuit optimization. This is particularly true because the JFET contributes a negligible amount of heat compared to the tube implementations.

A workhorse JFET crystal oscillator is shown in Fig. 2-30. With the capacitive divider and rf choke designated, this circuit will serve many communications system needs in the 1-5 MHz range. Note the diode shunted from gate to ground. Such an element was not seen in the otherwise-similar tube circuits. Whereas it may be acceptable to drive the grid of the electron tube into its positive region, analogous forward-bias of the gate-source diode in the JFET is not desirable. The tube continues to amplify when there is grid current, but the JFET does not operate properly when there is gate current. Gate current causes abrupt limiting of amplification; at best it causes harmonic generation—at worst, it can stop oscillation, especially if the crystal is not too active. Besides, the manufacturer's specifications does not advocate such operation. It turns out in practice that a small positive voltage at the gate is not detrimental, but means are needed to prevent heavy conduction. To this end, the 1N914 diode is used to divert most of the current that would otherwise circulate in the gate-source circuit of the FET.

The equal-valued capacitive dividers are generally satisfactory for this Colpitts circuit. (In bipolar transistor versions of this circuit, special attention is required for the capacitive divider beyond the mere compliance with feedback criteria. But, in the FET oscillator, there is much less need to swamp out or isolate the device's input impedance from the crystal.) The combination of the resistance and inductance in the source lead enables dc bias and high impedance from ground to be developed independently. At the same time, any stray resonances that might exist in the rf choke are prevented from being troublesome by the series resistance, which

+ 10 to 20 V

68

0.05 μF

1N914

1 Meg

68 pf

100 pf

Out

5 μH

68 pf

68

Suitable FETs include 2N5458, 2N5459, 2N4416, 2N3819

Fig. 2-30. FET oscillator for use with parallel-resonant 1-5 MHz crystals. This is a good workhorse oscillator combining simplicity with reliable operation. Many practical oscillators using specific semiconductor devices are discussed in Chapter 5.

acts as a Q spoiler. Although a 100 pf output coupling capacitor is shown, its size should be reduced to the minimum value consistent with the needs of the load. A small trimmer capacitor of, say 15 pf or so, can be connected across the crystal to provide a limited range of frequency variation.

The dc operating voltage and currents required by solid-state devices can exert considerable influence on the selection of such an element for an oscillator circuit. Although this is often merely a matter of convenience, the input and output impedances of oscillator devices depend upon *how* a given output power is produced. If low voltages and high currents are used, the impedances will be low. The converse is true for high dc voltages and low currents, in which case tube-like impedances may be simulated. When high frequencies and high-power levels are both required from an oscillator, the input and output impedances may trend awkwardly low and this would be aggravated by the use of a high-current low-voltage device. The trouble is that low impedances generally require low inductances and high capacitances in the resonant and impedance-matching circuits—nanohenries of inductance and tenths of microfarads of capacitance would, for example, be awkward or downright impractical to implement. Other things being equal, the power MOSFET displays higher impedance levels, at least at the input, than bipolar transistors.

Another factor in the selection of the semiconductor active device has to do with dc operating polarity. Whereas tubes were restricted to one polarity, bipolar transistors are available in their npn and pnp versions so that positive or negative power sources can be used. Similarly, MOSFETs are made in n-channel and p-channel fabrications. In either case, however, it must be ascertained that the device with the desired polarity is available with the other features needed for desired oscillator performance.

A factor sometimes overlooked in production runs of oscillator circuits is the wide tolerances generally pertaining to bipolar transistors. Whereas the amplification factor or transconductance of electron tubes may have been within a $\pm$ 10% variation, the current-gain factor (beta) of a given type of transistor may vary as much as three to one. An oscillator that performs well with a "hot" transistor may be reluctant to start with a transistor having a more typical beta value.

The logic devices usually operate in their digital modes and produce square wave outputs. The linear devices perform well in the oscillator circuits formerly used with tubes and operate in a similar fashion. Not shown are other solid-state devices that are useful in relaxation or negative resistance oscillators. Such devices would include the SCR, and PUT, the unijunction transistor (UJT), the tunnel diode, the lambda diode, the gunn diode, and the IMPATT diode. It can be seen that the designer, technician, and hobbyist now has a greatly expanded array of devices and circuits to consider.

Chapter 3

Theory of Oscillators

W E NOW EXAMINE THE OSCILLATION THEORY IN SOMEWHAT more detail than has been expedient during our discussions of oscillating elements and oscillation-provoking devices as separate entities of the oscillator. It is appropriate that we combine these basic oscillator components and investigate the overall effect. In particular, we will concern ourselves with feedback oscillators. All feedback oscillators present a negative resistance to the resonant tank circuit. The essential difference between those oscillators we specifically designate as of the negative-resistance variety and those we call feedback oscillators, involves a practical consideration that does not appreciably affect the theory of operation. In the negative resistance oscillator, the tank circuit can be separated from the device exhibiting negative resistance. In the feed-back oscillator, the oscillation provoking device no longer displays negative resistance if we sever it from the resonant tank. This is illustrated in Figs. 3-1(A) and (B), where the tank circuits are shown disconnected and replaced by a milliammeter. At Fig. 3-1(A) negative resistance is still present as indicated by a decrease in plate current as plate voltage is increased. At Fig. 3-1(B) an increase in plate voltage causes and increase in plate current, indicating positive resistance rather than negative.

THE TUNNEL DIODE

The tunnel diode is a two-terminal semiconductor device with

Fig. 3-1. Comparison of negative-resistance oscillator and feedback oscillator. The negative-resistance device displays negative-resistance with or without external feedback. (A) Negative-resistance oscillator. (B) Feedback oscillator.

direct capability of inducing oscillation in an associated tank circuit. From a practical viewpoint, this provides the most illuminating demonstration of the principle of negative resistance. The tunnel diode incorporates a pn junction and bears some similarity to junction rectifiers and zener diodes. In particular, its characteristics suggest a zener diode already in its avalanche condition without any externally provided bias. This being the case, a slight bias voltage supplied from an external source suffices to project the internal charge condition of the junction from zener or reverse conduction to forward conduction. In the accomplishment of this transition, an increase in bias voltage is accompanied by a decrease in bias current. Such a voltage-current relationship corresponds to negative resistance. The negative resistance thereby produced cancels a numerically equal amount of ordinary, or positive, resistance in resonant circuits. Under proper conditions, this results in sustained oscillation.

In other oscillation-provoking elements, the maximum frequency of operation is ultimately limited by transit time of whatever charge carries are involved. That is, the time required for

108

electrons, holes, or ions to traverse a distance between two electrodes establishes a limit beyond which higher frequency oscillation is not attainable even though the effects of stray capacitances and inductances are circumvented. In the tunnel diode, however, the transit time is infinitesimal with respect to generation of radio frequency energy. The high-frequency limit imposed by this factor is at least ten million megahertz. Thus, we find extensive usage of the tunnel diode in microwave techniques and radar applications.

The Two Oscillation Modes of the Tunnel Diode

When the tunnel diode is used in conjunction with reactive elements, two types of oscillation may be obtained. Either sinusoidal or relaxation oscillations will occur, depending very largely upon the regulation of the dc power source and its voltage. This is best illustrated by the test circuits of Fig. 3-2. In Fig. 3-2(A) a *low* value of limiting resistance is used. As the potentiometer is advanced, the "N" characteristic curve of the tunnel diode is followed. If the operating point were fixed in the center of the negative-resistance region, sinusoidal oscillations could be produced by properly as-

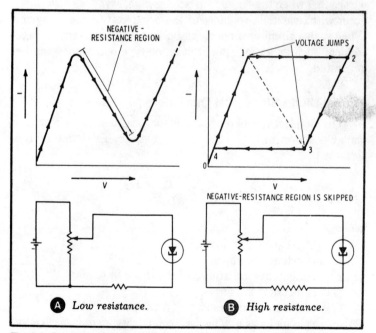

Fig. 3-2. Operational modes of tunnel diode corresponding to two types of oscillators.

sociated an LC tank with the circuit. This is so because the *nega-tive* resistance of the tunnel diode could then be used to effectively cancel the dissipative effect of ordinary or *positive*, resistance in the LC tank.

In the event the limiting resistance is *high*, that is, if the dc regulation of the power source is relatively poor, an entirely differ-ent situation exists. The is depicted in Fig. 3-2(B). As the poten-tiometer is advanced, the tunnel diode current increases along its characteristic from 0 to 1. Advancing the potentiometer further however, does *not* cause the operating point to trace its path along the downward slope of the negative-resistance region. Instead, an abrupt *jump* occurs with the operating point establishing itself at 2. Now, let the potentiometer be slowly returned toward its zero voltage position. The operating point will follow the characteristic downward from 2. After thus approaching point 3, the operating point will *not* locate itself on the slope of the negative-resistance region. Rather, a very slight further retardation of the potentiom-eter will cause an abrupt voltage *jump* from 3 to 4. If we now *ad-vance* the potentiometer, the 1, 2, 3, 4, cycle will be repeated. Thus, we see that when the series resistance is high, *switching transitions* occur with essential avoidance of the negative resistance region. When such a circuit is properly associated with an energy storage element such as an inductor, relaxation-type oscillations can be generated.

Tunnel-Diode Sinusoidal Oscillators

The simplest tunnel-diode oscillator for generating sinusoidal waves is shown in Fig. 3-3(A). The frequency f is found by the formula:

$$f = \frac{1}{2\pi} \sqrt{\frac{1 - (R \times 1/r)}{L \times C_1}}$$

where,
 f is the frequency in hertz,
 R is resistance in ohms,
 r is the negative resistance of the diode in ohms,
 L is inductance in henrys,
 C is capacitance in farads.
What this circuit lacks in schematic complexity is compensated for by the abundant, and often highly mathematical, technical litera-ture dealing with the theory of tunnel-diode operation. For the cir-

Fig. 3-3. Tunnel-diode sine-wave oscillators.

cuit of Fig. 3-3(A), a basic relationship establishes the nucleus for more extensive analysis of oscillatory action. The time constant given by the net positive resistance of the circuit in conjunction with the inductance must be equal to, or less than, the time constant due to the capacitance of the diode in conjunction with the negative resistance of the diode. Expressed symbolically, $L/R = rC_1$, where L is the inductance in the circuit; R is the net circuit resistance; r is the negative resistance of the diode; and C_1 is the effective capacitance of the diode.

In order to obtain approximately linear operation, that is, dynamic excursion of the operating point over the essentially straight portion of the negative-resistance region, L/R must be equal to, or less than rC_1. This is most readily brought about by adjustment of R. However, a disadvantage of this circuit is that this relationship is difficult to maintain by variation of any *single* parameter. In other words, if we desire sinusoidal waveshape, we cannot readily change frequency by the convenience of varying the inductance, L.

A more stable and useful circuit is shown in Fig. 3-3(B). The frequency of oscillation of this circuit is found by the formula:

$$f = \frac{1}{2\pi} \sqrt{\frac{1}{L\,(C + C_1)} - \frac{(1/r)^2}{C_1\,(C + C_1)}}$$

where,

    f is the frequency in hertz,

    r is the negative resistance of the diode in ohms,

    L is inductance in henrys,

    C and C_1 are capacitances in farads.

Here, the criterion of oscillation is described by a somewhat more complex relationship:

$$R = \frac{1/r}{(2\pi f)^2 + (1/r)^2}$$

where,
f is the frequency of oscillation.

In this circuit, the frequency can be readily tuned without causing the dynamic operating range to exceed the approximately linear segment of the negative-resistance region. In both circuits, R cannot *exceed* a certain value without causing a change in oscillatory mode. When R is too large, relaxation type oscillations are produced due to abrupt switching between operating points. Although R is depicted as a single series resistance, in practical oscillators R is generally the *net* resistance due to a voltage divider and the dc resistance of the inductance, L.

Push-Pull Tunnel-Diode Oscillator

For the purpose of approximately doubling power output, we may devise push-pull versions of the basic single-diode oscillators. The circuit of Fig. 3-4 is a particularly useful one. The tunnel diodes can be biased to operate in the relaxation, or switching mode. This tends to provide relatively high output and reliable oscillatory action. At the same time, the output can be a fairly good approach to a sinusoid. This is brought about by the combined effects of harmonic discrimination by the tuned tank, and the tendency of the push-pull configuration to cancel even-order harmonics.

Cascade Tunnel-Diode Oscillator

An alternate method of increasing output power is to connect simple tunnel-diode oscillators in series, at the same time modifying the dc bias network so that only one battery or other dc source is needed. This technique is not limited to two diodes; rather, any number can be so cascaded. A cascade arrangement involving two diodes is shown in Fig. 3-5 for comparison with the push-pull circuit of Fig. 3-4. It should be noted that the relative polarity of the diodes differs in the two circuits.

With unfortunately chosen parameters, the circuit of Fig. 3-5 can operate so that the phasing in the two tank sections does not induce additive emf's in the output winding. This operational mode

112

Fig. 3-4. Push-pull tunnel-diode oscillator and waveforms.

can be discouraged by the use of relative high ratios of C to L in the tank sections.

Whereas, in the push-pull circuit, the two diodes alternate switching states, in the cascade arrangement both diodes undergo their switching cycles simultaneously. That is, the switching cycles of the transistors are in phase. Biasing for linear mode operation is also feasible in this arrangement.

Several practical oscillators are shown in Fig. 3-6. In Fig. 3-6(A), the circuit performs as the series-parallel type. Because of the high frequency, the physical capacitor that ordinarily is shunted across the diode is not required. In this case, the internal capacitance of the diode serves the function of C1. The circuit in Fig. 3-6(B) is of the series-parallel type as shown in Fig. 3-3(B). It will be observed that all three circuits illustrated in Fig. 3-6 make use of voltage-divider networks rather than the single series resistance of the basic circuits of Fig. 3-3. This is because most practical power sources, certainly commercially available batteries, provide voltages far above the several hundred millivolts needed to bias tunnel diodes in the middle of their negative-resistance regions. Also, the

Fig. 3-5. Cascade tunnel-diode oscillator.

113

Fig. 3-6. Tunnel-diode oscillators.

circuit of Fig. 3-6(A) utilizes the so-called "back diode." This is a tunnel diode that operates under *reverse*, rather than forward, bias conditions. These diodes generally operate at lower power levels than forward-conduction types, but at the same time permit the use of inductors with readily obtainable values. It should be appreciated that R in the basic circuits of Fig. 3-3 is, in the general case, the *net dc* resistance due to voltage-divider networks, coil resistance, and internal resistance of the power source.

The crystal-controlled oscillator of Fig. 3-6(C) is unique, being suggestive of tube or transistor oscillators employing a crystal in the feedback loop rather than directly as in a conventional tank circuit. In this arrangement, oscillation would not exist without the crystal because the net ac resistance R would then exceed the criterion for sinusoidal oscillation. At the same time, the voltage is sufficiently divided down so that relaxation oscillation cannot take place. With the crystal in the circuit, the two resistances are effectively paralleled for the frequency corresponding to series resonance of the crystal. Thus, the net value of R is lowered, enabling oscillation to be sustained in the LC tank circuit.

THE CLASS-C FEEDBACK OSCILLATOR

Our emphasis on feedback theory will cast much light on the operating principles of the most frequently encountered oscillator type in practical electronics—the Class-C oscillator. Although the Class-C oscillator is, at least in terms of one viewpoint, a switching circuit that shock-excites an LC tank with narrow pulses, we will find that much insight can be gained by treating such an oscillator as an overdriven feedback oscillator. Thus we will, in many instances, consider the Class-C oscillator as a feedback oscillator in which operation extends above and below the linear portion of the transfer characteristics of the tube or transistor.

Defining the Feedback Oscillator

In previous discussion oscillators in general were described as being devices that automatically converted dc power to ac. We are now confining our interest to feedback oscillators in particular, a feedback oscillator being an amplifier that supplies its own input signal. It does this via a feedback loop. The feedback loop is a signal path that permits a portion of the output voltage (or current) to be reimpressed at the input of the amplifier. Generally, but not necessarily, the feedback loop is a circuit external to the amplifier proper. Simply returning a portion of the amplified output signal to the input of the amplifier is not in itself sufficient for oscillation; the phase, or polarity, of the returned signal must be the same as the input signal responsible for its appearance. In this way, the original signal is reinforced in amplitude. The reinforcement is cumulative; a given change in the input signal produces a change in the amplified output signal, which in turn assists the input signal in continuing the process. The theoretical buildup to infinite amplitude inferred by this process is, in practice, limited to a finite value. We have here a condition of regeneration, and the feedback is said to be positive. Oscillation occurs at a frequency determined by a resonant tank, which functionally becomes part of the feedback network. The components of the feedback network in feedback oscillators are shown in Figs. 3-7 and 3-8.

THE QUESTION OF ORIGINAL SIGNAL VOLTAGE

Before defining additional criteria for the onset of oscillation

115

OSCILLATION IS OBTAINABLE
AT AMPLIFIER OUTPUT

AMPLIFIER

AMPLIFIER INPUT

+β

1   2   3

FEEDBACK NETWORK

(1) PHASING PROVISION FOR POSITIVE FEEDBACK
(2) VOLTAGE (OR CURRENT) APPORTIONING CIRCUIT
(3) FREQUENCY DETERMINING MEDIUM

Fig. 3-7. Basic elements of the feedback oscillator.

in the generalized system, it is expedient that we resolve a possible conflict in describing oscillator operation. Technical authors generally assume an oscillator is already in oscillation, then proceed to relate the manner in which such oscillation is sustained. In so doing, the reader is sometimes left to wonder how the oscillations began. It is easy to see that an oscillator is an amplifier that supplies its own input signal, but we are tempted to ask, from

Appropriate phasing of
feedback-coil connections
establishes positive feedback

Mutual inductance provides
feedback path

Active device (universal-
amplifier) provides power
gain necessary to overcome
circuit losses

Resonant circuit determines
oscillation frequency

dc source provides energy
which supplies circuit
losses and sustains oscillation

Fig. 3-8. Function of the elements in a feedback oscillator.

whence came the output signal when the oscillator was first turned on. Surely, there could be no output signal without an input signal, and conversely no input signal without an output signal.

INITIATION OF OSCILLATION BUILDUP

A simple two-terminal element, such as a length of wire or a resistance, is not, strictly speaking, passive. Thermally excited electrons produce minute, but for some purposes significant, voltages across the ends of such elements. Specifically, we have a *random-frequency generator* that develops an effective rms voltage, V_N, defined by the equation:

$$V_N = \sqrt{V_0^2 + V_1^2 + V_2^2 + V_3^2 + V_\infty^2}$$

where we designate all frequencies from dc, V_0, to infinite frequency, V_∂. At any instant, any frequency has a statistically equal chance of occurring. That is, V_N is pure white noise.

In real life, a definite resistance associated with a finite bandwidth is involved, and so is a temperature in excess of absolute zero. Thus, it is found that V_N can be calculated as follows:

$$V_N = \sqrt{4k \times T \times BW \times R}$$

where,

V_N is the rms value of the noise voltage comprising all frequencies within bandwidth BW,

BW is in hertz,

T is the absolute temperature (centigrade plus 273°),

R is the resistance of the two-terminal element in ohms,

k is Boltzmann's constant (1.38×10^{-23}).

At room temperature of 21 °C, this equation simplifies to:

$$V_N = \sqrt{1.623 \times 10^{-20} \times BW \times R}$$

The presence of V_N at the input of a positive-feedback amplifier constitutes the "start" signal for initiation of a regenerative buildup process culminating in sustained oscillation at the frequency dictated by the frequency-determining elements contained in the feedback loop. Thus, we see that any practical amplifier or oscillator always has a "built-in" input signal.

The switching of power-supply voltage often induces transients in the resonant tanks that shock-excite oscillations, thereby providing the initial signal for cumulative buildup. In some cases, it would be difficult to ascertain whether power-supply transients or thermal-noise voltage was the more influential in starting the regenerative cycle leading to amplitude oscillation. In any event, a definite time is required for the buildup process. The higher the Q of the resonant tank, the greater is the buildup time. Buildup time also depends upon feedback, being shorter with heavy feedback. A crystal oscillator with just sufficient feedback for oscillation can consume several seconds in attaining constant level operation. Such an oscillator obviously would not be suitable for interruption by a telegraph key.

EFFECT OF FIXED BIAS ON
SPONTANEOUS OSCILLATION BUILDUP

A power oscillator may be provided with a source of fixed bias in addition to grid-leak bias. Such an arrangement is often desirable for large tubes as a protection against cessation of oscillation. We know that grid-leak bias exists only during oscillation. If for some reason such as excessive loading, oscillation should stop, a ruinous plate current could quickly destroy the tube, unless some other biasing scheme was incorporated to provide negative grid voltage in the absence of oscillation. However, if a fixed bias is too high, neither thermal noise voltage nor circuit transients would suffice to start the buildup process. Such a situation would, of course, exist if the fixed bias projected the tube into its cutoff region. The tube would provide no amplification under this condition. Sometimes an oscillator with such a bias arrangement is encountered in diathermy or high-frequency heating applications. Oscillation is started by momentarily shorting (through a relatively low resistor) the fixed-bias supply.

EFFECT OF POSITIVE
FEEDBACK ON GAIN OF AN AMPLIFIER

Because of the popularity of high-fidelity audio equipment, it is common knowledge that negative feedback imposes a reduction of amplifier gain, but permits certain desirable improvements such as extended frequency response, better stability, and relative independence from tube and circuit variations. Some oscillators use negative-feedback loops, and we shall have more to say regarding negative feedback later. However, our primary concern in oscillators is the nature of positive feedback. Let us designate the voltage gain of an amplifier without feedback by K, and its voltage gain with feedback by K'. We wish to establish a relationship between K and K'. We intuitively suspect that the amount of feedback returned from the amplifier output to the amplifier input will be involved. Let us then designate the fraction of the output voltage reapplied at the input by the symbol β. β will be plus for positive feedback (regeneration) and, appropriately, negative, for inverse feedback (degeneration). These quantities will then be related according to the equation:

$$K' = \frac{K}{1 - K\beta}$$

We see from this equation that positive feedback increases the gain of an amplifier.

PHYSICAL INTERPRETATION OF INFINITE GAIN

Further contemplation of the equation reveals that it is possible for the denominator to approach or even become zero. Such a condition means the gain of our amplifier becomes infinite. What, indeed, is the practical significance of infinite gain? Actually, before infinite gain is ever attained, the amplifier bursts into oscillation at a frequency that produces an in-phase relationship between input and feedback signals. The in-phase relationship, in turn, is governed by the sum of all reactive components (inductors and capacitors) in the amplifier circuit. We see why a regenerative detector, being essentially an amplifier with controllable positive feedback, permits such high amplification from a single amplifier. In Fig. 3-9 we see a simple amplifier circuit with several modifications added to make the relationship

$$K' = \frac{K}{1 - K\beta}$$

more meaningful with respect to our study of oscillators. In this formula, K is the gain without feedback, and K' is the gain with feedback. We see that β, the portion of output voltage fed back to the input, is determined by a voltage divider. The transformer is connected to invert the phase of the amplified signal appearing in the circuit. Familiarization with this basic circuit will provide a great deal of insight into the operating principles of feedback os-

$$= \frac{R_y}{R_x + R_y}$$

Source isolating resistance

ac source

V_{IN}

V_{OUT}

ac voltmeters

Feedback switch

R_x

R_y

1:1 audio transformer connected for negative feedback

Fig. 3-9. Test circuit for study of feedback. Feedback equations can be confirmed with the input and output ac voltages measured in this negative-feedback amplifier.

cillators, notwithstanding the complexity of the feedback network or the novelty of various approaches in other circuitry sections of the oscillator.

An interesting feature of feedback oscillators is shown in Fig. 3-10. Here we see a graph depicting the minimum value of β required for oscillation as a function of K, the nonfeedback voltage gain of the amplifier. Significantly, the graph illustrates that, when K is high, β is relatively low. Usually, it is desirable to have K high from the standpoint of frequency stability. This is particularly true in tube circuits when the frequency-determining tank is in the grid circuit, for then the plate and load circuits will be better isolated from the grid circuit than would be the case with low K and the necessarily high β.

FEEDBACK AND NEGATIVE RESISTANCE FROM THE "VIEWPOINT" OF THE RESONANT TANK.

We have seen that oscillation in a feedback oscillator occurs when the product $K\beta$ is equal to one. This gives us the mathematical condition of infinite amplification in the equation relating amplification with feedback, K', to amplification without feedback, K. We should now reflect upon the physical significance of equating

Fig. 3-10. Maximum feedback signal required for oscillation.

$K\beta$ to unity. This condition signifies that we have apportioned just sufficient output voltage to exactly compensate the feedback network losses. Generally, the main component of the feedback network is the resonant tank. Consequently, the tank has applied to it a signal ($K\beta$) of sufficient amplitude to compensate for the damping action of the losses, which would otherwise cause oscillations to die out in the event they were somehow started. We see that the basic philosophy involving action on the resonant tank is not qualitatively different than that occurring in the negative-resistance oscillator. In both the negative-resistance oscillator and the feedback oscillator, the tank circuit becomes oscillatory when its equivalent resistance is matched by a numerically equal negative resistance, the negative resistance being a generator or amplifier condition that cancels the effect of ordinary losses, or "positive" resistance. The equivalent resistance, R_0, of a parallel-tuned tank is $L/R_S C$. In the dynatron oscillator then, the dynamic plate resistance of the tube, $-r_p$, must be equal to the quantity $L/R_S C$ for oscillation to take place. It is likewise true that the resonant tank of the feedback oscillator "sees" $L/R_S C$ ohms of negative resistance when feedback conditions are just right to sustain oscillation (this relationship is pictured in the diagram of Fig. 3-11).

Basic Consideration of Phase of Feedback Signal

It has been mentioned, but without emphasis on phase conditions, that the product of $K\beta$ must be at least unity for the condition of oscillation to be attained in the feedback oscillator. It is, however, also necessary that the fractional output signal, β,

Fig. 3-11. The mathematical principle of oscillation reconciles the differences between feedback and negative-resistance techniques.

returned to the input be in-phase with the input signal. A vacuum-tube amplifier of conventional circuit configuration (grounded cathode) causes a phase displacement of 180° between output and input signals. A 180° phase shift corresponds to polarity inversion; when the input signal is undergoing a positive excursion in its cycle, the amplified output signal is at a corresponding negative point in its cycle. We see that direct feedback from plate to grid must produce negative, not positive, feedback; that is, β is negative. When we insert a negative value of β in the expression $K/(1 - K\beta)$, the resulting amplification given by this expression shows a decrease, not the necessary increase, or specially the mathematical identity with infinite gain. We see the need for a means of shifting the phase of the output signal 180° before returning it to the input. The way in which this is accomplished in practice largely accounts for the many different circuits of feedback oscillators. We know some of the ways wherein such a phase shift can be obtained. Transformers, autotransformers, series-resonant tanks, additional amplifier stages, and RC networks can be used for this purpose. (In the event that a tube is used as a grounded-grid amplifier, no phase shifting is required in the feedback network to bring about the condition of positive feedback necessary to produce oscillation.)

Feedback Under Various Phase Conditions

We have discussed the condition of positive feedback, and have pointed out the opposite nature of negative feedback. What about the phase conditions that correspond neither to ideal in-phase or out-of-phase signal return? It is only by investigating this more general condition that we can gain a comprehensive insight into the basic concept of oscillation in the feedback amplifier. This purpose is best served by studying feedback conditions in an amplifier with a feedback loop over a wide frequency range. We may even suppose that the amplifier is of the so-called "negative feedback" type in which positive feedback is not deliberately designed into the circuit. By measurement or computation of the amplitude and relative frequency range, it will always be found that negative feedback departs more and more from the ideal out-of-phase relationships as frequencies much lower or much higher than the design bandwidth are checked. Indeed, at certain low and high frequencies, it will be found that the feedback becomes positive. This is a natural consequence of the total circuit reactances. Actual capacitors and inductors tend to bring about this condition at low frequencies, whereas at high frequencies stray capacitance and transformer leakage inductance are more generally involved. If, at a frequency corresponding to exact in-phase positive feed-back, the loop gain $K\beta$ equals or exceeds unity, the amplifier becomes oscillatory at that frequency.

The foregoing principles are illustrated in Fig. 3-12. Positive feedback is pictured at Fig. 3-12(A), with the feedback signal exactly in phase with the original signal. Oscillation will result if the amplitude of the feedback signal is sufficient, that is, if $K\beta = 1$. At Fig. 3-12(B), an ideal negative-feedback condition is shown, with the feedback signal exactly 180 degrees out of phase with the original signal. At C there is positive feedback not exactly in phase with the original signal. Oscillation cannot occur even for large amplitudes of the feedback signal. Fig. 3-12(D) shows negative feedback not exactly 180 degrees out of phase and Fig. 3-12(E) shows six different phases of feedback. At phase P6 there is an exact in-phase relationship, and oscillation will take place if $K\beta = 1$.

THE PRACTICAL OBSTACLE TO INFINITE BUILDUP

One of the basic operating features of all oscillators is the slowing down and ultimate cessation of the oscillatory buildup rate. We intuitively know that this happens, for otherwise the oscillatory amplitude would theoretically increase forever; tremendous circulating currents would exist even in a physically small tank circuit, and the process would come to a halt only as a result of a burnt-out inductor or capacitor. There is nothing in the nature of the resonant tank that otherwise tends to restrict the amplitude buildup of oscillations when the effect of its losses are neutralized by feedback or negative resistance. Therefore, we must investigate the

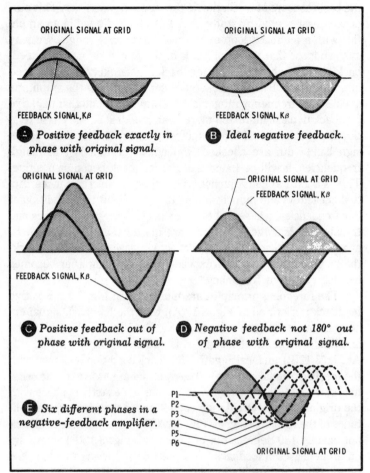

Fig. 3-12. Various feedback conditions in amplifiers and oscillators.

oscillation-provoking device. In so doing, we find that all such devices, no matter how divergent their operating principles, possess one common feature. As circuit elements they all show non-linear relationships between voltage and current, provided voltage or current is either very small or very large, relative to the values within the center of more ordinary operating regions. For example, the Class-A tube type amplifier, which we almost instinctively associate with linear or proportional signal transfer, is linear only when the ac grid and plate voltages are small enough to limit operation over straight-line portions of the graphs representing grid and plate voltage-current relationships. Let us consider several specific instances in which nonlinearity exerts a braking effect upon oscillation buildup.

AMPLITUDE LIMITING BY AUTOMATIC
BIAS CONTROL IN CLASS-C OSCILLATORS

In the Class-C oscillator, it is most common to employ automatic bias. The input of the amplifying device is biased towards cutoff by the oscillation itself. Before such an oscillator commences operation, no negative input-bias is developed and signal amplification, K, is at, or near its maximum value. Assuming that the value of $K/(1 - K\beta)$ is, or exceeds, unity, oscillation buildup will start. The amplifying device then draws current through the automatic bias resistance ("gridleak" from tube-technology) and develops a voltage drop across this resistance. As a consequence of the rectifying action of the active device's input circuit and the filtering action of the input capacitor, a negative input bias is developed. The greater this bias, the greater is the reduction in the amplification. This is shown in Fig. 3-13(A). It is seen that sufficient negative-bias will bring the amplifier to the point where no further buildup of oscillation amplitude can take place.

From B of Fig. 3-13 it can be seen that oscillation amplitude initially builds up rapidly, but soon its rate of increase slows down.

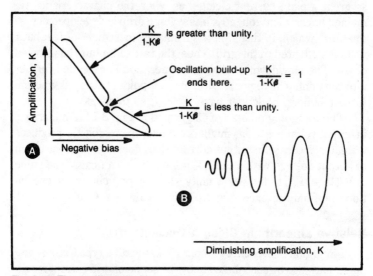

Fig. 3-13. The amplitude-limiting action of automatic bias in an oscillator. (A) Amplification as a function of the negative input bias. (B) Amplitude limiting caused by oscillation-produced bias.

The increase comes to a halt when sufficient negative bias is developed so that $K/(1 - K\beta)$ is unity. Thus, the Class-C oscillator, because of oscillation dependent bias, has a built-in amplitude-limiting mechanism.

Amplitude Limiting in Class-A Oscillators

In the Class-A oscillator, we do not have the automatic amplitude-limiting action of grid-leak bias. Here, the nonlinearity in the plate-voltage plate-current relationship exerts the limiting function. A tube operating in a Class-A oscillator is a voltage controlled device; a grid-voltage change produces a plate-current change, which by its flow through an impedance inserted in the plate circuit develops a voltage change of greater magnitude than the original grid signal. We see that although a voltage is fed back to the grid in an oscillator, the voltage is the result of plate current. Consequently, the ability of the tube to produce plate-current changes in response to grid-voltage changes is directly relevant to feedback oscillator operation.

Suppose oscillation is building up and has attained an amplitude corresponding to operation beyond that straight-line relationship between plate voltage and plate current (see Fig. 3-14). With further buildup, a given increment of grid signal no longer produces as great a plate current change, as when the characteristics remained linear. Consequently, less voltage drop is developed across the tank, which, in turn, implies a smaller increment of feedback voltage delivered to the grid. Thus, the rate of buildup has slowed down. The slowdown continues as operation is extended further into the nonlinear regions. Ultimately, a point is reached wherein further buildup no longer pays its way, so to speak.

The foregoing points are diagramed in Fig. 3-14. An increase in output voltage, such as might occur over the region XY produces a smaller increase in output current than does a similar variation of output voltage within the XX region. This can cause the value of $K/(1 - K\beta)$ to fall below unity, thereby preventing the oscillation amplitude to exceed approximately the YY level.

Relative Linearity in Class-A Oscillators

Limitation of oscillatory amplitude in the tube-type Class-A amplifier may also be contributed to by the action of grid current resulting from excessive feedback. Although little negative bias results therefrom, the load imposed by the current-consuming grid on the

126

Fig. 3-14. Limiting action in a Class-A oscillator. Because all active devices depart from linearity at low and high operating levels, equal increments of voltage produce equal increments of current only over a restricted amplitude range.

feedback coil or network lowers the grid excitation, and thereby the oscillatory level in the tank circuit. The objection may be raised that the definition of Class-A amplification restricts operation to linear grid and plate regions. However, the typical Class-A oscillator is provided with relatively low feedback, just enough to project operation into nonlinear regions. Operation remains essentially Class-A. In the Class-A amplifier, despite some departure from true linearity, ac grid voltage and ac plate current are very nearly sinusoidal. The output waveform is therefore relatively free from harmonics, even before being acted upon by the filtering action of the resonant LC circuit.

Amplitude Limiting in Negative-Resistance Oscillators

In the dynatron negative-resistance oscillator, it might appear that unrestricted oscillatory buildup would take place because the amplifying function of the tube is not used. Here, we might postulate that a decrease in gain is not meaningful inasmuch as the tube behaves directly as a negative resistance. Although such an argument is sound with respect to the scope of its implication, it falls short of accounting for the overall principle of operation. In much the same way that the would-be inventor of a perpetual motion machine always finds success blocked by the manifestation of friction in some form, just so is our dynatron oscillator destined to limit oscillatory buildup via nonlinearity. The nonlinearity involved is more apparent than in the feedback oscillator, for it appears in the characteristic curve. The same is true of the transitron oscillator, the point-contact transistor negative-resistance oscillator, and other nonfeedback oscillators which supply negative resistance to the resonant tank circuit. Inspection of the negative-resistance region of the current versus voltage graph of a negative-resistance device shows that a linear, or near linear, region of negative resistance exists over a reverse slope region such as XX in Fig. 3-15. When oscillatory amplitude builds up a little bit beyond the level corresponding to operation within region XX, positive resistance is encountered, and the buildup rate is slowed down. Equilibrium is soon attained, for the positive resistance brings the circuit to its criterion of oscillation. Further buildup, with its accompanying heavier contribution of positive resistance, cannot occur because the circuit would then no longer satisfy the criterion of oscillation. The criterion of oscillation is $-r_p = L/R_sC$, where $-r_p$ is the negative resistance of the device and L/R_sC is the equivalent positive resistance of the resonant, parallel-tuned LC circuit. It is interesting to contemplate that the same criterion holds true in a feedback oscillator, that is, at the same time $K/(1 - K\beta) = 1$, $-r_p = L/R_sC$.

NEGATIVE-RESISTANCE DEVICE $-r_p$

CRITERION OF OSCILLATION : $-r_p \cdot \dfrac{L}{(R_S)(C)}$

$R_S \cdot R_L + R_C$

A *Circuit.*

POSITIVE-RESISTANCE REGIONS

CURRENT

VOLTAGE

B *Limiting action.*

Fig. 3-15. Limiting in a negative-resistance oscillator.

(In this case, $-r_p$ is interpreted as the negative resistance "seen" by the resonant tank.)

DIVERGENT EFFECTS OF GRID BIAS IN FEEDBACK AND NEGATIVE-RESISTANCE OSCILLATORS

Although an appreciation of features common to both feedback and negative-resistance oscillators cannot fail to enrich our comprehension of the oscillatory process, it would tend to defeat our purpose if we overlooked significant differences between these basic types of oscillators. We have already commented upon the fact that severance of the tank circuit destroys the negative-resistance property in the feedback oscillator, whereas negative resistance is displayed by the tube in a negative-resistance oscillator independently of the presence or characteristics of the tank. An interesting operating difference is found with regard to the response of the two oscillator types to grid bias. For simplicity, we will compare the Class-A feedback oscillator with the dynatron negative-resistance oscillator. In the first circuit, increasing the value of negative bias makes oscillation *less likely*. We know this is so because the gain (K) of the tube is thereby reduced. In the dynatron, however, oscillation with a given tank circuit may not take place until the grid bias is made sufficiently negative. How do we account for this contradiction in behavior? For a given plate voltage, more negative grid bias produces decreased plate current. This is equivalent to saying that the plate resistance of the tube has been made higher; this statement applies to both static and dynamic plate resistance. Not only is the plate resistance higher in the positive regions of the plate characteristics, but more important; it is also higher in the negative-resistance region. Thus, the ability of the dynatron to "reach" the equivalent positive resistance of a resonant tank is enhanced as grid bias is made higher. Too low a negative grid bias causes $-r_p$ to be numerically higher. Too low a negative grid bias causes $-r_p$ to be numerically less than L/R_sC, and oscillation cannot occur. The same reasoning applies to the transitron if we interpret $-r_p$ as representing the negative resistance of the dynamic screen-grid characteristic.

THE MULTIVIBRATOR

The multivibrator (Fig. 3-16) is a feedback oscillator for generating relation-type oscillations, that is, pulses or nonsinusoidal waves. Basically, the multivibrator consists of a two-stage resis-

Fig. 3-16(A). The multivibrator relaxation oscillator. The pulse repetition rate is approximately R1C1 + R3C2 Hz.

$$\frac{1}{R1C1 + R3C2}$$

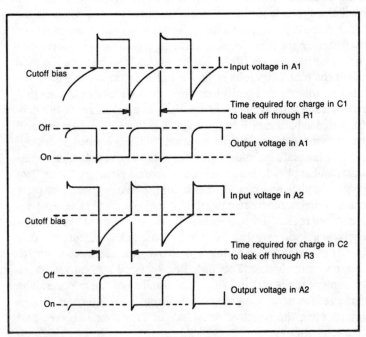

Fig. 3-16(B). Waveforms in multivibrator oscillator.

tance capacitance-coupled amplifier with a feedback path between output and input. The theory of operation is best approached by arbitrarily considering one stage as an ordinary amplifier and the other as a phase inverter even though their circuitries are exactly

130

the same. Assume, also, that both stages are grossly overdriven so that they operate in both, their saturated, and their cut-off regions.

Suppose a positive pulse from thermally-generated noise appears at the input of "amplifier" A1. The momentary increase in output current in A1 then couples a negative-going pulse to the input of "phase-inverter" A2. The amplified, but inverted pulse appearing in the output circuit of A2 is fed back to the input circuit of A1, the polarity (positive) being proper to reinforce regeneratively the original noise transient. A rapid and cummulative buildup now takes place in the circuit until abruptly stopped by cut-off in the output current of A2. When A2 is driven to cutoff, A1 is driven to a state of heavy conduction. This circuitry condition persists until the charge accumulated in capacitor C2 can leak off through R3.

When the voltage across C2 is sufficiently depleted, stage A2 again conducts; in so doing, it produces a negative-going pulse at the input of stage A1. Now, A1 is cutoff and remains so until the accumulated charge in capacitor C1 leaks off through R1. When this happens, stage A1 reverts to its conductive state, thereby delivering a negative pulse to A1, cutting off conduction in this stage. It is seen that the two stages are alternately on and off, the rate being determined by RC time constants. Each switching action is regenerative; once initiated in one stage, the alternate stage helps drive it to completion. Such regenerative action produces steep-sided rectangular waves.

THE BLOCKING OSCILLATOR

The blocking oscillator, like the multivibrator, produces relaxation-type oscillations, which have a pulse-repetition rate governed by the time-constant of an RC network. Unlike the multivibrator, the blocking oscillator contains an LC tank circuit. It is seen in Fig. 3-17 that the tank circuit consists of the effective inductance of the input winding of transformer T, in conjunction with distributed and stray capacitance. The schematic configuration of the oscillator is essentially that of a simple tickler-feedback circuit. However, the two windings of transformer T are very closely coupled and the time constant of the input capacitor C and resistance R is much larger than in an oscillator intended for continuous oscillation at, or near, the resonant frequency of the tank circuit; that is, C is a larger capacitor and/or R is a higher resistance than would be employed in sinusoidal feedback oscillators.

Fig. 3-17. Basic operation of the blocking oscillator. (A) Circuit. (B) Waveforms.

Let us investigate how these differences alter the operation of the oscillator.

Action in Blocking Oscillator

Assume that the buildup in an oscillation cycle has begun. Output current is increasing and due to the phasing of transformer T, the input circuit of the amplifier is being rapidly driven positive, that is, into conduction. The two actions are regenerative, with one reinforcing the other. These actions proceed until output current saturation abruptly halts further increase. When this occurs, the positive drive to the input of the amplifier can maintain itself only for a relatively brief duration determined by the energy storage of the transformer. This is so because a transformer secondary develops induced voltage only when the primary current is changing in value. When the positive voltage at the amplifier-input end of the transformer finally collapses, the amplifier input is left with a high negative bias.

The development of this bias was previously discussed in the discourse on Class-C oscillators and amplifiers. Briefly, the conduction of the forward-biased (positive) input of the amplifier charges capacitor C so that its amplifier-connected plate is negative with respect of its transformer-connected plate. During the charging process, the trapped charge is overcome and cannot manifest itself as negative bias. However, shortly after the amplifier is driven into saturation, the charging current disappears, leaving the amplifier input biased with a sufficiently negative voltage to cutoff the amplifier's output current. That is, a rapid transition has been made from the conductive to the nonconductive state of the amplifier. The amplifier is maintained in its *off* state until the charge trapped in capacitor C has had time to sufficiently deplete itself through resistance R. With a large capacitor and a high resistance, the quiescent time of the blocking oscillator may be many times the duration of the generated pulses.

The sudden cessation of output current causes generation of a half-cycle of overshoot in the output winding of the transformer. The blocking oscillator is essentially a self-switching circuit that combines the operating principles of the inductively-coupled (tickler) feedback oscillator and the multivibrator. The pulse duration is determined by the natural oscillation frequency of the effective tank-circuit. On the other hand, the pulse repetition rate is determined by the RC time constant of capacitor C and resistance R. It is important to keep in mind that charge time of capacitor C is very much less than discharge time. This is because the charging resistance, composed of the transformer secondary and the forward resistance of the amplifier input, is much lower than the discharge resistance provided by resistance R.

THE SQUEGGING OSCILLATOR

The squegging oscillator is similar in principle to the blocking oscillator. However, instead of a single pulse being generated, discrete wavetrains of high-frequency oscillation are produced at periodically-recurrent intervals (see Fig. 3-18(B)). Any oscillator circuit operating in Class-C becomes a squegging oscillator when input capacitor C and/or resistance R is made so high that the charge produced in C does not have time to decay appreciably between

Fig. 3-18. The squegging oscillator. (A) Circuit. (B) Waveform.

high-frequency oscillatory cycles. The amplifier, under such conditions oscillates in spurts.

There is little difference between this type of oscillator and the blocking oscillator. The blocking oscillator generally employs an iron or ferrite-core transformer, resulting in very strong feedback. In addition, the effective Q of the equivalent tank is relatively low. This design approach results in quick disablement of amplifier operation. Conversely, in the squegging oscillator, conditions are such that many cycles of high-frequency oscillation are required to attain amplifier cutoff.

SINE WAVE OSCILLATION IN
THE PHASE-SHIFT OSCILLATOR

The first network we will consider consists of three-cascaded high-pass networks. Now, at first glance, such a composite circuit may not appear significantly different than the two-similar RC networks used in the multivibrator, assuming that the separation of the networks by an active-device is of no consequence in the comparison. The important difference between the two and the three-stage RC networks is that a 180-degree phase-shift is readily obtainable from the three-stage network, but not from the two-stage network. This being the case, we see that such a three-stage network can be used for phase inversion in returning a signal (of correct phase for positive feedback) from the output to the input of a *single* active-device. The circuit of such a "phase-shift" oscillator is shown in Fig. 3-19. If additionally, the active device gain is sufficient to satisfy the criterion of oscillation, $K/(1 - K\beta = 1$, oscillation must occur.

Inasmuch as no resonant LC-circuit is present, what will be the frequency of oscillation? The oscillation frequency will be that at which the three-stage RC network produces a 180-degree shift in phase. Unfortunately, the attenuation of such a network is very high, that is, the factor β in $K/(1 - K\beta)$ is low. This implies the need for an active-device with appreciable gain. It can be mathematically shown that transistors must develop a current gain of 29. FETs, op-amps, and tubes must develop a voltage gain of 29. Because of circuit losses, such as those due to the low-impedance input of transistors, gains well in excess of 29 are often required in practical oscillators.

Good wave-purity is attainable from phase-shift oscillators because harmonics do not produce a 180-degree phase-shift, and therefore cannot be supported in the oscillatory process. Phase-shift oscillators find greatest use at audio frequencies. With large, low-leakage capacitors, they are able to generate good sine-waves at very-low frequencies. When a transistor is used, a good rule-of-thumb is to use 10 kilohm resistors for all three shunt-arms of the RC network, and for the collector load, as well. Although certain

Fig. 3-19. The phase-shift oscillator. Unlike multivibrator-circuits employing RC networks, the phase-shift oscillator delivers a pure sine-wave. In order for oscillation to take place, the voltage-gain of the amplifier must be at least 29.

frequencies and capacitor availabilities may not allow this, it will usually be found to be a good starting point.

THE PARALLEL-T OSCILLATOR

The parallel-T network also produces a 180-degree phase-shift at a "resonant" frequency. An oscillator using this network is shown in Fig. 3-20. This network additionally develops maximum attenuation coincident with the phase inversion. In other words, it behaves as a frequency-rejection filter. This attenuation must be compensated by adequate gain in the active device. It fortunately happens that the attenuation of harmonic frequencies is even greater than that accorded by the network to the fundamental, i.e., the oscillation frequency. This is because the phase-shift of the harmonics is *not* 180 degrees. This implies varying amounts of negative feedback, thus reducing the amplification of these harmonics. The practical ramification of this is that the parallel-T oscillator is also capable of exceptionally-good wave purity.

Although vernier changes in frequency can be brought about by changing any single network element slightly about its design value, an extensive tuning-range requires the ganging together of

Fig. 3-20. The parallel-T oscillator. Also known as the Twin-T oscillator, this circuit is much used in low and audio-frequency applications. It is capable of exceptionally good wave purity.

either the three-resistances or the three-capacitances in the network.

Op-amps tend to be the best active-devices for implementing this type of oscillator. Transistor designs are most likely to be successful if *two* transistors are used—a first-stage as a conventional common-emitter circuit and a second-stage as an emitter-follower circuit. The external feedback path is then made from the output of the emitter-follower to the input of the T-network.

THE WIEN BRIDGE OSCILLATOR

This is probably the most widely exploited type of RC feedback oscillator for sine waves. The name is somewhat of a misnomer, for the elements enclosed in the positive feedback loop do not make up a complete Wien bridge. The "half-bridge" (R1C1, R2C2) involved in the positive feedback loop displays radically different characteristics than are associated with a complete Wien bridge. The response curves for a full bridge and a half bridge were previously shown in Chapter 1. Whereas proper employment of the full bridge yields a nulling network or frequency rejection filter, the half-bridge in the positive feedback loop of the Wien bridge oscillator produces a maximum response at the "resonant" frequency. Furthermore, the half-bridge gives us zero, rather than the 180-degree phase-shift associated with resonance in the complete Wien bridge. In the circuit of Fig. 3-21 the remaining two arms

Fig. 3-21. Basic Wien bridge oscillator. When properly designed, excellent sine waves are produced by this circuit.

137

of the bridge are used in a negative feedback loop and are not required for oscillation. Their function in the performance of the oscillator will be discussed later.

Inasmuch as the half-bridge in the positive feedback loop does not invert the phase of the fed-back signal, an additional amplifying stage is provided for this purpose. It is seen that each of the two amplifying stages produces a 180-degree of phase shift. The total phase shift is therefore the needed 360 degrees. This in-phase relationship exists for only the one frequency determined by the resistances and capacitances in the half-bridge arm of the positive feedback loop. At other frequencies, this network produces a phase shift other than 360 degrees or zero. It can be appreciated that this leads to sine-wave generation—loop phase shift is not right for support of harmonics. Nonetheless, more has to be added to this oscillator for high quality sine waves.

The negative feedback loop is provided so that operation can be limited to the linear regions of the amplifiers. This leads to excellent wave purity. Resistance R_k is generally a tungsten filament lamp, or other thermosensitive element with a positive coefficient of resistance (resistance increases with temperature, and therefore with current). Because of its circuit position, this resistance provides self-bias to stage A1. At the same time, this resistance is the shunt arm of a voltage divider consisting of itself in conjunction with series arm R_f. This network acts as a variable voltage divider in the negative feedback-loop of the oscillator. The feedback is negative because the output from stage A2 is returned to the noninvert input of stage A1.

Suppose that the oscillation amplitude is building up. The alternating current through R_k is increasing as a result. In turn, the resistance of R_k is increasing, which causes a much higher proportion of negative feedback to be developed. A point is ultimately reached wherein the net positive feedback in the circuit is just sufficient to maintain the amplitude level. A higher buildup in amplitude cannot occur because the proportionally much greater increment of negative feedback would reduce the amplification below that needed to overcome circuit losses and maintain oscillation.

If, on the other hand, the oscillator output tends to decrease for any reason, the proportionally much greater increment of negative feedback removed from the circuit by the cooling of the lamp would permit sufficient increase in net positive feedback to very nearly restore the amplitude to its previous level.

The thermal inertia of the lamp is great enough to prevent cyclic heating and cooling above, say 100 Hz. At lower frequencies some distortion is produced due to the fact that the lamp resistance begins to follow the ac wave. Tuning is accomplished by ganging two elements of the halfbridge in the positive feedback loop, usually the two capacitors. The frequency is inversely proportional to the capacitance, rather than to the square-root of capacitance as in LC oscillator. Consequently, a given change in tuning capacitance enables coverage of a much greater frequency ranger than in LC oscillators.

LOADING OF OSCILLATORS

The loading of an oscillator is a very important consideration. When *power is extracted* from a resonant tank by means of inductive or capacitive coupling to the load, a *reduction in* Q always occurs. This is because the load resistance is "reflected" into the resonant LC circuit as an equivalent resistance which increases the effective value of R_S. In the interest of frequency stability, it is desirable to maintain a high operating Q of the tank. This, in itself, imposes the desirability to limit the amount of loading to a value *less* than that corresponding to maximum power in the load. There are other factors which also make such a compromise worthy of consideration. The oscillator, being a self-excited device, needs some minimum grid excitation to maintain oscillation. When we load an oscillator, the resultant reduction in tank-circuit Q makes it more difficult to generate sustained oscillation. Unfortunately, the extraction of power from the tank also deprives the grid of excitation. This, in conjunction with the lowering of the Q, will cause oscillation to stop if loading is carried too far. Even before oscillation is stopped, the operation becomes unstable, with increased harmonic generation and very high plate current. Unless economic or other factors do not permit, it is always preferable to employ a lightly loaded oscillator in conjunction with a tuned power amplifier. With such an arrangement, maximum power may be delivered to the load by the amplifier with relatively little reaction on the oscillator.

Detuning Effect Due to Loading; the Faraday Shield

Another difficulty encountered with the loading of an oscillator is the detuning effect of reactance in the load circuit. Such reactance is coupled into the resonant LC circuit in a similar manner to the resistive component of the load. At radio frequencies, it is practically impossible to escape this effect. In some instances, it is convenient to retune the tank circuit to return the oscillator to its original frequency; in other cases such a procedure would constitute an annoyance, at the very least. A load-coupling coil should always be located near the "cold," that is, the ac ground, end of the output resonant circuit; the "hot" end of the LC tank is susceptible to *capacitive tuning* by the proximity of the metal in the coupling coil. The *Faraday shield* shown in Fig. 3-22, represents an additional technique to reduce this effect. A Faraday shield consists of a grid structure of parallel wires connected by a single contacting wire that is grounded. The Faraday shield is inserted between the load-coupling coil and the output resonant tank. Of course, this cannot isolate capacitance in the actual load circuit. A Faraday shield must not contain any closed loops, inasmuch as it is intended to act upon electrostatic lines of force but must be "transparent" for electromagnetic induction. The transference of power from the oscillator tank to the load, by means of a capacitor, is another loading method. The detuning effect is generally more pronounced than in the case of inductive coupling. This can often be considerably overcome by using a large capacitor tapped down on the inductor of the output tank circuit, rather than a small capacitor directly connected to the "hot" end of the tank. Inasmuch as capacitive reactance becomes smaller with higher frequency,

Fig. 3-22. The Faraday shield.

$$R_x = (L1/L2)^2 \, (M2)(R)$$

Fig. 3-23. Loading characteristics of an LC feedback oscillator. (A) Circuit: R_x is load reflected into LC tank. (B) Dc currents in oscillator. Oscillator no longer is operative in shaded region because of overloading and load power falls to zero.

capacitive coupling can be expected to transfer more harmonic power to the load than electromagnetic coupling.

Plate Current Is Function of Effective Tank Circuit Q

The plate current necessary to maintain oscillation in a Class-C feedback oscillator using an LC tank circuit as the frequency determining medium is approximately $1/Q$ times the alternating current in the inductive or capacitive branch of the tank. The no-load plate current of an oscillator with a high-Q tank is lower therefore than when a low-Q tank is used. By the same reasoning, we anticipate an increase in plate current when an oscillator is delivering power to a load, for loading decreases the operating Q of the tank. The loading characteristics of an LC feedback oscillator are shown in Fig. 3-23.

The Need for Load Isolation

Let us suppose that we use a Faraday shield between oscillator tank and load coil and that the reactive component of the load is cancelled by tuning, that is by insertion of an equal reactance but of opposite kind. We would still find that the addition of the load affected the frequency of the oscillator. Why is this? We have, for simplicity, been using the resonance formula,

$$f_0 = \frac{1}{2\pi\sqrt{LC}}$$

For most of our discussion, this is justifiable when O_0 is ten or greater. Inasmuch as oscillation frequencies often have to be maintained within relatively tight tolerances, it is well to appreciate that the mathematically rigorous relationship is:

$$f_0 = \left(\frac{1}{2\pi\sqrt{LC}}\right)\left(\sqrt{\frac{L - R_L{}^2 C}{L - R_C{}^2 C}}\right)$$

We see that the resonant frequency of a tank is affected by resistance as well as by the more influential inductance and capacitance. Thus, the resistance reflected into the tank by coupling the load to it changes the frequency of the oscillator. The resistance reflected into the tank also lowers its Q. The desired oscillation frequency can be restored by a tuning adjustment of the tank capacitor or inductor. The degradation in Q cannot be remedied for a given amount of power extracted from the oscillator. The decreased Q makes the oscillator frequency less stable with respect to changes in tube characteristics, changes in operating voltages, the effects of temperature variations, and mechanical disturbances. What is needed, we see, is a transfer of power from oscillator to load of such nature that the parameters of the load could not react upon the resonant tank. Such unilateral power transfer exists in the electron-coupled oscillator illustrated in Fig. 3-24.

THE ELECTRON-COUPLED OSCILLATOR

The electron-coupled oscillator uses a tetrode or pentode tube in an arrangement in which the plate of the tube is *not* involved in the generation of oscillation. Rather, a complete oscillator circuit is formed with the cathode, control grid, and screen grid. These elements act respectively as the cathode, grid, and plate of an ordinary triode oscillator tube. We know that the screen grid in a tetrode or pentode tube intercepts a portion of the accelerated electrons from the cathode and that the remainder complete their journey to the plate. Those that reach the plate in the electron-coupled oscillator are velocity modulated at the oscillation frequency. This results in similar plate-current variations. Thus, ac power at the oscillation frequency is available in the plate circuit. When this power is dissipated in a load, a unilateral power transfer prevails; power is delivered from the screen grid (the "plate" of the oscillator proper) to the tube plate, and the load circuit, but the load cannot appreciably influence the oscillator circuit. Neither the condition of maximum power of transfer, nor partial power transfer into a reactive load can exert much detuning, or Q-degrading, influence on the oscillator tank circuit. Some power amplification occurs between screen grid and plate. In this respect, the screen grid acts as a control grid of a triode amplifier. Indeed, the electron-coupled oscillator displays many of the features of an oscillator-amplifier arrangement. The screen grid is at ac ground potential. Therefore, it acts as an electrostatic shield, further isolating the oscillator and load circuits.

Fig. 3-24. Basic electron-coupled oscillator and curves showing frequency variation.

Circuitry Considerations in the Electron-Coupled Oscillator

The electron-coupled oscillator shown in Fig. 3-24 uses a Hartley oscillator in the cathode/control-grid/screen-grid circuit. The Hartley oscillator has not been discussed in this section. In an endeavor to explain concepts basic to all feedback oscillators, we have confined our examples to simple "tickler" type feedback circuits. We see, however, that the Hartley configuration bears a similarity to the tickler feedback circuit. Essentially, a tapped coil rather than two separate coils is used. It is not the Hartley circuit in itself that is necessary to the electron-coupled oscillator. A tickler feedback circuit could be devised as well. The Hartley circuit as shown has the advantage that the screen grid is readily operated at ac ground potential. In the technical literature we also find electron-coupled oscillators with shunt-fed Hartley circuits in which the screen grid is "hot." This type of electron-coupled oscillator is satisfactory at low frequencies. Above several hundred kHz there may exist excessive capacitive coupling between the screen grid and plate, thereby sacrificing the high degree of isolation attainable when the screen grid is operated at ac ground potential.

In order to avoid confusion, our electron-coupled oscillator is shown with a choke in its plate circuit. Sometimes a transformer, or, in very low power applications, a resistor, is used. Improved waveform purity results when a resonant tank tuned to the oscillation frequency is used. When this is done, it must be ascertained that the output is not due in part, or in entirety, to the "tuned-plate tuned-grid" oscillator formed by having resonant tanks in the input and output circuits. Such malperformance is more probable at high than low frequencies, and is also more likely with tubes providing insufficient screen-grid shielding action.

Additional Factors in the Performance of the Electron-Coupled Oscillator

When a pentode is used in the electron-coupled oscillator, it is generally best to ground the suppressor grid. This provides an additional Faraday shield between oscillator and load circuits. When the suppressor grid is grounded, there is less objection to an oscillator circuit in which the screen grid is above ac ground potential. When properly designed, the electron-coupled oscillator overcomes yet another trouble attendant to the ordinary self-excited oscillator. A change in dc plate voltage generally produces a shift in oscillation frequency in these circuits. In the electron-coupled

144

oscillator, however, the screen-grid voltage and plate voltage changes resulting from a given change in dc power-supply voltage tend to shift the oscillation frequency in opposite directions. A wise choice in the ratio of plate to screen-grid voltage can result in high immunity to frequency change with respect to changes in the dc operating voltage from the power supply. The power output from the electron-coupled oscillator is limited for two reasons. First, the screen grid does not have sufficient power-handling capability to drive the "amplifier" portion of the tube to high output. Secondly, the Class-C bias requirements of the *oscillator* section impose a limit on the output from the amplifier section. We see that the oscillator section functions as an electron gun for the amplifier section, but the operation of both sections is very much determined by the automatic bias developed across the grid leak in the control-grid circuit.

Chapter 4

Practical Oscillators

IN THIS CHAPTER WE WILL EXAMINE THE CHARACTERISTICS OF some of the basic feedback oscillator circuits. In so doing, we shall find ourselves in the advantageous position of being able to draw freely upon the fundamental concepts discussed in the preceding chapters.

Although oscillators such as the Hartley, the Colpitts, the tuned-plate/tuned-grid, and others can all be represented by a block diagram depicting the basic function of amplification, phase reversal, frequency determination, and feedback apportioning, a close scrutiny of these oscillators reveals distinctly individual "personalities." It will now be our concern where the resonant tank is located, the actual nature of the feedback path, the means whereby the returned signal is caused to be in phase with the input signal, and other details which have been deemed of sufficient importance to merit the formal naming of specific oscillator circuitries. Hitherto, we have confined our discussions to "tickler" feedback circuits in which inductive coupling between grid and plate coils in tube circuits established the feedback action. This type of oscillator sufficed as a model for analyzing performance parameters common to all feedback oscillators. We will now proceed to other specific circuitries of the same family, i.e., LC feedback oscillators and closely related variants, such as crystal oscillators.

THREE TYPES OF HARTLEY OSCILLATORS

The Hartley oscillator is one of the "classical" LC feedback circuits. Unfortunately, the technical literature abounds with confusion with respect to the basic principles involved in the operation of this circuit. There are actually *three* mechanisms whereby oscillation may be produced in a Hartley oscillator. Although all three oscillatory modes are generally present, it will serve our objective best to consider each one separately. In this way, we will gain practical insights into this "sure-fire" oscillator. The derivation or evolution of the Hartley oscillator from the basic "tickler" feedback oscillator is depicted in Fig. 4-1. We will designate a Hartley oscillator with mutual-inductance tank-inductor sections, L1 and L2, as type 1. A Hartley circuit without such coupling between L1 and L2 will be dealt with as type 2. Finally, we shall consider as type 3, a configuration similar to that of type 2, but with such small internal capacitance in the transistor that there is no chance of oscillation as the result of feedback through such internal capacity. It will be interesting to see whether such a "stripped down" Hartley circuit can still oscillate.

The Type-1 Hartley Oscillator

From Fig. 4-1 it is evident that the Type-1 Hartley is, indeed, very similar to the tickler-feedback circuit. What has been done is to substitute an autotransformer arrangement in place of a conventional transformer in the feedback circuit. In both circuits feedback occurs because of inductive or electromagnetic coupling between the output and input sections of the oscillator. It is differences of this nature that endow the various oscillator circuits with unique "personalities." To put the matter in more objective terms, some oscillator circuits are easier to implement in certain practical situations. In any event, the two Hartley circuits of Figs. 4-1(D) and 4-1(E) are generally considered easier to build and place in operation than the tickler circuits. (A tapped-coil tends to be less monkey-business than two separate coils, and it is easier to get a high value for M, that is, tight coupling.) Summing up, the Type-1 Hartley oscillates because of electromagnetic coupling between output and input.

The Type-2 Hartley Oscillator

Let us next consider the Type-2 Hartley circuit in which there is *no* mutual-inductance between the output and the input. As shown in Fig. 4-2, mutual-inductance, M, is zero. In practice, the coils could be far away from each other, or better, could be mounted so one

Fig. 4-1. Evolution of Type-1 Hartley oscillators from basic tickler feedback circuit. (A) Tuned base tickler oscillator. (B) Tuned base tickler oscillator with shunt dc feed. (C) Circuit (B) with physical rearrangement of tickler coil. (D) Shunt-fed Hartley (hot collector) oscillator. (E) Series-fed Hartley (hot emitter) oscillator.

is perpendicular to the other. Although the Hartley oscillator is not commonly constructed to attain this objective, it frequently happens that the coupling between L1 and L2 section of the tank is not as strong as it might be. How, then, can oscillation occur? One way in which the oscillator can still operate is via feedback provided by the internal collector-base capacitance of the transistor.

It may be argued that simply providing a signal path from collector to base results in negative, rather than positive feedback—it does not, in itself, produce the conditions required for oscillation. Recall, however, from Chapter 1, that an inductive load in the output circuit of an active-device can result in a negative input-resistance. It was also pointed out that a parallel-tuned tank circuit appears inductive when it is operated below its resonant frequency. We now see that our Type-2 Hartley circuit can oscillate even though there is no electromagnetic coupling between its output and input! This is so because at a frequency slightly below that corresponding to resonance of the tank circuit, the collector will "see" an inductive load. And the base circuit will then present a negative resistance to its tapped portion of the tank, thereby provoking it into oscillation.

The Type-3 Hartley Oscillator

In the Type-3 Hartley oscillator, we postulate that, like the type-2 circuit, there is no electromagnetic coupling between output and input; we further assume there is insufficient internal capacitance in the transistor to provide a feedback path. This implies that the negative-resistance mechanism described for the type-2 circuit will not be present. (Of course, if the Type-3 Hartley *is* capa-

Internal capacitance provides feedback path

L1

M = ZERO

L2

RFC

+

Fig. 4-2. The Type-2 Hartley oscillator. Oscillation takes place despite the fact that there is no inductive coupling between the input and output circuits.

ble of oscillation, the tank circuit will still "see" a negative resistance.)

The Type-3 Hartley circuit can be implemented with a tetrode or pentode tube, in which the internal capacitance is greatly reduced. However, we can simulate the operation under this condition by using the Type-2 transistor Hartley at a very-low frequency, say several hundred Hz. Such an arrangement is shown in Fig. 4-3. This circuit, can, indeed, be made to oscillate! The explanation is that the tank circuit now comprises a phase-shifting network—output energy is returned to the input, but displaced 180 degrees. With sufficient amplifier-gain, such an arrangement will oscillate, as witness the various phase-shift oscillators in which this condition is deliberately brought about via RC networks which are more lossy than LC circuits.

The Oscillatory Conditions
Generally Found in Practical Hartley Circuits

Practical Hartley oscillators tend to operate with all three of the described oscillatory modes. That may well explain the reputation enjoyed by the Hartley oscillator as being a "sure-fire" circuit. It has been also found that Hartley oscillators tend to have

Fig. 4-3. The Type-3 Hartley oscillator. Schematically the same as Fig. 4-2 this Hartley circuit, because of the low oscillation frequency, does not have sufficient collector-base capacitance to cause oscillation. And despite the lack of inductive coupling between output and input, this circuit *does* oscillate!

Fig. 4-4. Basic Lampkin oscillator. The objective in this circuit is to attain an approximate impedance match between the input of the amplifier and the resonant tank.

very good loading characteristics, and have served well from very-low to very-high frequencies. The tank inductor may be center-tapped, but this much excitation is not usually needed. Somewhat more stable operation results from locating the tap closer to the ground or "cold" end of the tank inductor.

THE LAMPKIN OSCILLATOR

The Lampkin oscillator is shown in Fig. 4-4. We see that the circuit is similar to the Hartley configuration, the essential difference being that the grid connection is tapped down from the capacitor-inductor junction of the tank. This minor change, can, however, result in a considerable improvement in frequency stability. In Hartley, as well as other feedback oscillators, which are commonly operated in the Class-C region, that is with grid current, the input impedance of the tube is generally much lower than the impedance of the tank circuit connected to it. (A notable exception is the Clapp version of the Hartley.) As a consequence, the tank circuit impedance is shunted down and its Q is thereby lowered. In the Lampkin circuit, the basic idea is to drive the grid-cathode circuit from a portion of the tank that constitutes an approximate impedance match. Of course, the input impedance of the tube varies over the oscillation cycle, so the impedance matching concept cannot be considered in a strict mathematical sense. Nevertheless, we can say that the average value of the input impedance during grid conduction is more closely matched by this technique, than by simply driving the grid from the "top" of the tank circuit. Insofar as concerns frequency stability and preservation of the Q of the resonant tank, it is acceptable that the tap be located low enough down the tank so that the grid is driven from a source having a lower impedance than that of the tube input circuit. (The same objective, that of isolating the tank from the tube, is basic to the Meissner oscillator and is evident in other LC oscillators where the grid or plate connection is tapped down from the capacitor-inductor junction of the tank. In the Clapp oscillator, freedom from the shunting effect of the tube is achieved by employing a series-resonant tank which inherently has a low impedance.) Although high mutual induction between the tapped portions of the tank inductor is not of primary importance in the ordinary Hartley, it is mandatory that close coupling exist between the inductor sections in Lampkin circuit. Otherwise, there will be a strong tendency toward oscillation at other than the desired frequency.

THE TUNED-PLATE/TUNED-GRID OSCILLATOR

Another classical LC feedback circuit stems from the Type-2 Hartley. If we tune the two tapped sections of the tank separately rather than employ one capacitor across the total tank turns, we arrive at the well-known tuned-plate/tuned-grid circuit (see Fig. 4-5). In the practical construction of this oscillator, care is exercised to prevent inductive coupling between plate and grid tank circuits. Oscillation then occurs as described in the Type-2 Hartley. The plate tank operates at a frequency slightly below its resonant frequency, thereby presenting an inductive component to the plate circuit. If a high-Q grid tank is connected to the grid-cathode circuit, oscillation takes place at very nearly the resonant frequency of the grid tank. In this oscillator, the tank circuit with the highest Q tends to exert the preponderance of frequency control. Although the tuned-plate/tuned-grid oscillator has traditionally been classified as a feedback oscillator, it would not be amiss to consider it as a negative-resistance oscillator. This, of course, is a subtle point. Any controversy provoked by such designation bespeaks well of the proponents of either argument. One of the central themes of this book is to stimulate thought concerning feedback and negative resistance. It is inevitable that insight into oscillator theory will

Fig. 4-5. The tuned-plate/
tuned-grid oscillator.

result from speculations regarding the equivalences and differences
between feedback and negative resistance. In any event, the tuned-
plate/tuned-grid oscillator is not as persistent an oscillator as is the
Hartley. However, due to the presence of two separate tuned circuits, this circuit tends to produce better wave purity than the Hartley. A well-screened tetrode or pentode will not work in this circuit
without a small physical capacitor connected from plate to grid.
When this is done, both plate and grid circuits must be made to
appear inductive, in order to form a phase-inversion network. The
action is not quite the same as that in a triode, but the difference
is more academic than practical.

THE MILLER OSCILLATOR

The Miller crystal oscillator (Fig. 4-6) is a tuned-plate/tuned-
grid oscillator, in which a piezoelectric crystal is substituted for the
grid tank. The crystal operates very close to its parallel-resonant
frequency. When a triode tube is used, oscillation occurs when the
plate load appears inductive inasmuch as the grid-cathode input circuit then presents a negative resistance to the crystal. When a well-
screened tetrode or pentode is used, it is sometimes necessary to
connect a "gimmick," a very small capacitance, across the plate
and grid terminals. The operation is then, for most practical purposes, identical with that obtained from a triode. The frequency
will be very slightly lower, for in this way both grid and plate tanks
operate slightly below resonance. By so doing, the plate and grid
inductances, with the capacitance connected from plate to grid, form
a phase-inversion network.

At communications frequencies, most small pentode tubes will
oscillate without the added external plate-grid capacitance. For

153

TUNE TO LOWER FREQUENCY THAN
PARALLEL RESONANT
FREQUENCY OF CRYSTAL

OUTPUT

Fig. 4-6. The Miller crystal oscillator.

given plate output levels, the pentode is to be preferred over the triode because the crystal is subjected to much less power. Both fracture and burning occur in crystals when the excitation is too high. In most applications where frequency control is the objective, crystal oscillators are operated at low power, on the order of a fraction of a watt to several watts, and drive an intermediate level (buffer) amplifier, or directly excite the power output amplifier. When the Miller crystal oscillator is designed around a junction transistor, it is often necessary to select a high activity crystal, due to the drastic impedance mismatch between the parallel-resonant crystal and the input of the transistor. In other respects, operation can be compared to its vacuum-tube counterpart.

THE COLPITTS OSCILLATOR

The Colpitts oscillator is the third of the "classical" LC feedback circuits. The two common versions of the Colpitts oscillator are shown in Figs. 4-7 and 4-8. The configuration of this oscillator resembles that of the Hartley. The difference is that the tapped connection is made at *capacitor* junctions rather than at *inductor* junctions. As a result, the tank circuit forms a phase-inverting network. The Colpitts cannot operate in an analogous way to the Type-1 Hartley because the electrostatic equivalent of electromagnetic induction is not present. Operation is most like that occurring in the Type-3 Hartley wherein the phase-inverting properties of the LC network make the most direct contribution to oscillation. The Colpitts circuit generates a frequency slightly higher than the resonance of the tank circuit. It is, indeed, under such a condition that phase inversion occurs between the output and input circuits of the active device.

The two capacitors constitute a *voltage divider* and their ratio governs the excitation of the oscillator. Sometimes we find these capacitors ganged, in which case constant excitation can be maintained over a fairly wide frequency range. Another tuning method

Fig. 4-7. The shunt-fed Colpitts oscillator. The capacitive divider formed by C1 and C2 is the hallmark of the Colpitts circuit. The quality of the rf chokes is important in the shunt-fed version.

that can attain similar ends is to employ fixed capacitors for C1 and C2 and a variable capacitor across the entire coil. This method is most satisfactory when the desired frequency tuning-range is moderate. For relatively-small tuning excursions, a vernier capacitor across either C1 or C2 is satisfactory.

The salient feature of the Colpitts is its comparatively good wave purity. This is due to the fact that C1 and C2 provide low-impedance paths for harmonics, effectively shorting them insofar as concerns the input of the active device. Another noteworthy feature of the Colpitts is that it is an exceptionally fine performer at high frequencies, well into the microwave region. (The active device must, of course, have high-frequency capability.)

Fig. 4-8. The Colpitts oscillator (series feed).

Fig. 4-9. Colpitts oscillator with crystal substituted for parallel-tuned tank.

Another type of Colpitts oscillator, with a crystal substituted for the inductor of the parallel-tuned tank, is depicted in Fig. 4-9. (In order to avoid possible undesired oscillation via other operating modes, the two chokes should not be identical.) In this circuit, the crystal operates at slightly below its parallel-resonant frequency and thereby simulates an inductance.

THE ULTRA-AUDION OSCILLATOR

Although schematically different, the ultra-audion oscillator is, in principle, a Colpitts oscillator. In Fig. 4-10 we see the circuit as actually wired, and the equivalent circuit taking into account the interelectrode capacitances of the tube. In particular, the grid to cathode capacitance and the plate to cathode capacitance form the familiar capacitive voltage divider of the Colpitts circuit. The ultra-audion is particularly well adapted for very high and ultra-high frequencies. The ultra-audion is basically a triode oscillator circuit. However, the less than perfect shielding provided by the screen grid of most tetrodes and pentodes, in conjunction with their high power gain, generally permits good operation with these tubes, too, particularly at higher frequencies. Inasmuch as it is difficult to make rf chokes with "broadband" characteristics at high frequencies, the ultra-audion local oscillator in many TV sets uses composition resistors in place of the chokes.

THE PIERCE OSCILLATOR

The Pierce oscillator is a novel circuit in that no LC resonant tanks are employed. We see in Fig. 4-11 that a crystal is connected from plate to grid. At a frequency slightly higher than that corresponding to the series-resonant mode of the crystal, the crystal will

Fig. 4-10. The ultra-audion oscillator.

appear inductive. This simulated inductance forms, in conjunction with the interelectrode capacitances of the tube, a Colpitts tank network (see Fig. 4-12). It might be argued that a similar circuitry condition would exist at a frequency slightly lower than the parallel-resonant frequency of the crystal. This is true, but the equivalent impedance in the vicinity of the series-resonant mode is much less than in the neighborhood of parallel resonance. As a consequence, feedback is greater near series resonance and the tube finds this mode more favorable for oscillation. Often it is found necessary to connect a small capacitor to the grid and cathode terminals of the tube to bring about oscillation. When such a capacitor is variable, a vernier tuning range is obtained. Sometimes a parallel-tuned tank is inserted in the plate circuit in place of the choke in order to improve wave purity. When this is done, the tank is operated at a frequency slightly higher than that corresponding to its resonance, in order to present a capacitive component to the plate-cathode circuit. Such an operating condition is automatically attained by simply tuning for maximum output.

THE CLAPP OSCILLATOR

The Clapp oscillator is another Colpitts derivative. The distin-

Fig. 4-11. The Pierce crystal oscillator.

Fig. 4-12. Equivalent circuit of the Pierce crystal oscillator.

L- CRYSTAL AT FREQUENCY SLIGHTLY HIGHER THAN
THAT CORRESPONDING TO SERIES RESONANCE.
C1 - CATHODE-GRID CAPACITANCE
C2 - CATHODE-PLATE CAPACITANCE

guishing feature of this circuit is that a series-tuned tank is employed as part of the total feedback network (see Fig. 4-13). Oscillation occurs at a frequency slightly higher than the resonant frequency of the series-tuned tank. Under this condition, the series-tuned tank appears inductive. Thus, an LC phase inversion network is formed in conjunction with voltage-divider capacitors C1 and C2 which is of the same basic configuration as the conventional Colpitts tank circuit. Due to the fact that the series-tuned portion of the tank has a much lower impedance than parallel-tuned tanks, it is postulated that the Clapp oscillator tank circuit is relatively immune to variations in tube parameters and loading conditions. This, it is contended by those favoring this oscillator, naturally manifests itself as excellent frequency stability.

There is much truth in the above assertion. However, other oscillator circuits also exhibit good performance when care is taken to limit excitation, when high-Q tanks are employed, and when temperature-compensating capacitors are inserted in the resonant circuits. The Clapp oscillator invariably embodies these precautions, and often incorporates electron-coupled output. This is because this oscillator has been employed primarily for governing the frequency of amateur transmitters. It history has involved very little in the way of workhorse applications. Conversely, the other oscillator circuits, primarily because of their readiness to oscillate, have often been designed and constructed in ways not conducive to optimum frequency stability. Although there is no magic shortcut to frequency stability in the Clapp oscillator, it does appear to be relatively easy to construct and place into satisfactory operation.

THE TRI-TET OSCILLATOR

The tri-tet oscillator (Fig. 4-14) is a circuit particularly well suited for the generation of selected harmonics of a crystal's fun-

Fig. 4-13. Clapp oscillator.

damental oscillatory frequency. Such frequency multiplication is often desirable because quartz crystals tend to become mechanically fragile, electrically inactive, and expensive at frequencies above, say, 10 MHz. An additional advantage of harmonic operation is that harmonically related communications frequencies may

Fig. 4-14. The tri-tet oscillator.

be readily selected; these harmonics have the same percentage frequency stability as the fundamental frequency. The name "tri-tet" derives from the simultaneous operation of the circuit as a triode oscillator and a tetrode amplifier. In this respect the tri-tet circuit resembles the electron-coupled oscillator. However isolation between oscillator and amplifier portions of the circuit is dependent upon the fact that the amplifier is tuned to a harmonic of the oscillation frequency.

The isolation between load and oscillator, which would be otherwise obtained from electron coupling, is negated by the common impedance, L1C1, inserted in the cathode lead. The oscillator section of the tri-tet circuit is a Type-3 Hartley in which the crystal acts as one portion of the total tank inductance and the parallel-tuned tank, L1C1, constitutes the other. In order for this to be so, the oscillation frequency is slightly below the parallel-resonant frequency of the crystal. For the same reason the oscillation frequency must be below the parallel-resonant frequency of tank L1C1. Indeed, increased harmonic output results when this tank is actually resonated to the desired harmonic. Output tank L2C2 is always resonated at the desired harmonic frequency.

When the output tank, L2C2, is tuned for fundamental frequency operation, the crystal current is likely to become dangerously high. This danger can be eliminated in two ways. First, tank L1C1 can be tuned to a very much higher frequency than the fundamental. In actual practice, L1C1 could remain tuned to one of the harmonics. Secondly, tank L1C1 can be short circuited. The circuit will then operate as a Miller or tuned-plate/tuned-grid oscillator because at radio frequencies there is usually sufficient internal feedback within the tube. If a well-screened tube is used at low and medium radio frequencies, it may be necessary to connect a "gimmick" capacitance between plate and grid terminals for fundamental frequency oscillation when tank L1C1 is short circuited. The harmonic generating ability of the tri-tet oscillator is directly related to its excitation; that is, it must operate deeply in the Class-C region in order that grid and plate current pulses are steep and of short duration (rich in harmonic content). This condition is enhanced by using an active crystal, and a high quality rf choke in the grid circuit. It should be appreciated however that a so-called "harmonic" crystal is not necessary in this circuit.

THE MEISSNER OSCILLATOR

The Meissner oscillator employs electromagnetic feedback, but there is no conductive connection between the resonant tank and the remainder of the circuit. In Fig. 4-15 there is no intentional mutual induction between L1 and L2. (In practical Meissner oscillators, the coupling between these coils is, at least, so loose that no oscillation is thereby produced.) There is, however, mutual induction between the inductor of the resonant tank and these coils. Oscillation occurs at approximately the resonant frequency of the tank, because feedback then attains its maximum value. The coupling between the three circuits should be no greater than that which is necessary for oscillation with a moderate load. (The load can be inductively coupled to L3.) Otherwise, this circuit is likely to exhibit a pronounced tendency to generate spurious frequencies. This is so because stray and distributed capacitances associated with L1 and L2 resonate these coils at frequencies other than those corresponding to tank-circuit resonances; there is always transformer-type coupling between the three coils regardless of the tuning of C3. It is highly desirable to use a Faraday shield between the tank coil, L3, and the grid and plate coils, L1 and L2. This greatly reduces the possibility of oscillation at spurious frequencies due to capacitive feedback. This oscillator is uniquely suited for certain instrumentation and measurement techniques, due to the fact that the resonant tank merely has to be brought in proximity with the electronic circuit proper. Under laboratory conditions, the Meissner oscillator may be adjusted to operate so that there is a relatively high isolation between tube parameters and the tank circuit. For practical applications, the Meissner oscillator is not as popular as the other LC feedback circuits.

Fig. 4-15. The Meissner oscillator.

THE MEACHAM BRIDGE OSCILLATOR

The Meacham bridge oscillator consists of a high-gain "flat" amplifier, in conjunction with a Wheatstone bridge containing a resonant arm (see Fig. 4-16). The bridge is inserted in a positive feedback loop between the output and input of the amplifier. The resonant arm of the bridge is generally a crystal. When the crystal is operated at its series-resonant mode, its effective resistance then tends to balance the bridge, thereby producing extreme (theoretically infinite) attenuation of the feedback signal. Under such a condition, oscillation could not exist. It is clear that a slight imbalance of the bridge is necessary to enable the circuit to meet the criterion of oscillation. It is not desired that this imbalance be attained by off-resonant operation of the crystal, for the phase shift thereby produced would again defeat the requirements of the oscillation criterion. If the phase were corrected elsewhere in the circuit, such operation would not advantageously make use of the optimum frequency stability that can be provided by the crystal. Bridge imbalance is attained by using a small tungsten filament lamp as one of the bridge arms. Such a lamp has a positive temperature coefficient; its resistance increases with operating temperature.

Let us suppose that the lamp is cold and the circuit has just been turned on. The bridge is grossly out of balance, enabling a

Fig. 4-16. The Meacham bridge oscillator. Operating very close to balance, the bridge automatically regulates feedback, thereby maintaining Class-A amplifier operation, exceptional frequency stability, and good wave purity.

strong feedback signal to be impressed at the input of the first amplifier stage. The resulting oscillatory current heats the lamp filament; the attendant increase in lamp resistance causes the bridge to approach its balanced condition. Feedback is accordingly reduced until equilibrium is attained with the bridge imbalanced just sufficiently to permit oscillation. Any tendency for the oscillatory amplitude to change is counteracted by a change in lamp resistance, which shifts bridge balance in the direction required to oppose the change. We see that the bridge not only governs oscillation frequency, but also provides amplitude stabilization. The amplifier operates in its Class-A region, for in the Meacham bridge oscillator the performance objective is frequency stability, not power. Other things being equal, frequency stability is generally enhanced by good wave purity and by amplitude stabilization. An outstanding feature of the Meacham bridge oscillator is that the bridge circuit operates with an effective Q several times higher than that of the crystal itself. This is due to the fact that a near-balanced Wheatstone bridge displays a phase-magnifying property.

In Fig. 4-16 we see "plus-beta" symbols. It should not be inferred from this that there are two positive feedback paths. A feedback path in any feedback oscillator involves a complete circuit, or loop. Most often, one portion of such a loop is the grounded part of the circuit. This makes it schematically convenient to simply designate the "hot" connection, or connections, with the plus-beta symbol, it being then understood that the ground side of the feedback loop also participates in the feedback of energy from the output of the oscillator to the input.

In the basic Meacham oscillator of Fig. 4-16 neither side of the external feedback path is grounded. The two plus-beta symbols merely indicate the route whereby occurs one complete go and return path, or loop for the return of energy from the oscillator output to its input circuit.

In the event one of the two connections designated by plus-beta signs in Fig. 4-16 were grounded, the basic principle of oscillation would not be altered. In such a case, it would be conventional to label only the ungrounded feedback connection with the plus-beta sign. Whether or not such grounding would be employed in a practical Meacham oscillator, would depend very much upon the kind of electrostatic shielding used in the bridge components and the associated transformers. The basic idea here would be to avoid imbalancing the bridge in such a way that the frequency-governing effect of the crystal would be impaired. In most instances, this matter would be of little consequence for frequencies up to the middle audio range of the spectrum.

LINE OSCILLATORS

Line oscillators are advantageously used at frequencies sufficiently high to enable practical exploitation of the resonant properties of transmission lines. A guidepost here is that the wavelength corresponding to a frequency of three hundred megahertz is one meter. A quarter-wave line for this frequency would therefore be approximately nine inches in length. Resonant lines associated with practical uhf oscillators are always physically shorter than the theoretical free-space wavelength. In the first place, the velocity of electromagnetic wave propagation along the line is slightly less than in free space. This is a relatively small effect in designing line oscillators, however. The major shortening effect is produced by the action of tube and stray capacitances across the high-impedance end of the resonant line. In Fig. 4-17 we see a simple line oscillator, which resembles the low-frequency version (using lumped tank circuit) of a push-pull ultra-audion oscillator. The exact oscillator type is often obscure in such uhf oscillators because there is sufficient "radiation coupling" between the resonant lines and the tube elements. Uhf output energy is extracted from the cold end of the resonant line by means of a half turn "hairpin" loop electromagnetically coupled to the line. The impedance of such a loop is low enough to feed directly into coaxial cable. Vernier tuning of this oscillator is provided by a small trimmer capacitor connected near the shorted end of the line.

Fig. 4-17. A line oscillator for uhf.

164

Fig. 4-18. Line oscillator employing high-Q quarter-wave coaxial line for frequency control.

Another Line Oscillator

Another line oscillator is shown in Fig. 4-18. Basically a tuned-plate/tuned-grid circuit, frequency control is provided by a quarter-wave section of coaxial line. The inductive reactance needed in the plate circuit for oscillation results from operation of the conventional "lumped" LC tank slightly below parallel resonance. Somewhat improved operation may result from the use of a butterfly resonator in the plate circuit (a tuning unit in which inductance an capacitance are integrally incorporated in a single mechanical assembly). However, frequency control is primarily exerted by the quarter-wave coaxial line in the grid circuit. Such a line may be constructed to have an extremely high Q. Theoretically Q is maximum when the ratio of outer conductor diameter to inner conductor diameter is approximately 3.6 to 1. Practically, ratios between three and five appear to be quite satisfactory. Temperature effects may be overcome by constructing the line of alloys such as Invar or Monel metal which have a low coefficient of expansion. In order to prevent degradation of the high Q inherent in such a structure, the inner conductor should be tapped as close to the shorted end of the line as is consistent with oscillation. The coaxial line is inherently shielded and does not radiate as does the open type two-conductor line. This avoids a dissipative loss and also makes the resonant frequency immune to proximity effects of nearby objects.

THE MAGNETOSTRICTION OSCILLATOR

The magnetostriction oscillator makes use of electromechanical resonance of magnetostrictive rods as described in Chapter 1.

It will be seen from Fig. 4-19 that the circuit configuration is suggestive of the Hartley oscillator. However, the phasing of the two windings is opposite to that required to produce oscillation in the Hartley circuit. As a result, there is sufficient negative feedback to discourage Hartley type oscillation. Oscillation occurs at the *mechanical resonant frequency* of the magnetostrictive rod, for only then is voltage phased for positive feedback induced in the grid-cathode winding. An important aspect of the circuit of Fig. 4-19 is that dc plate current flows in one winding. This establishes an operating bias about which oscillation occurs. The oscillation frequency is:

$$\frac{\text{speed of sound in rod}}{2 \times \text{length of rod}}$$

The same units of length should be used in both numerator and denominator of this relationship. If the speed of sound is designated as so many unit lengths per second, the frequency is obtained as so many hertz. For example, the speed of sound in a certain nickel alloy may be of the order of 4800 meters per second. A rod one meter in length would then oscillate at the frequency of 2400 hertz.

Fig. 4-19. The magnetostrictive oscillator. Despite the topographical resemblance to a Hartley circuit, the oscillation frequency is governed by the magnetostrictive rod not by LC relationships. Also, the phasing of the windings is such as to prevent Hartley oscillation.

Fig. 4-20. Basic Franklin oscillator.

If the circuit were arranged so that no magnetic bias was provided, the rod would tend to vibrate at twice the frequency of the exciting current because contraction results from electromagnetic forces of both polarities. Such operation is not very efficient. In order to avoid damping the effective Q of the vibrating rod, it must be clamped at its nodal point. This is the point, or narrow region, where vibrational amplitude is minimum. Another technique relevant to increasing the Q is to use a laminated bundle of thin strips of material rather than a solid rod.

THE FRANKLIN OSCILLATOR

The Franklin oscillator uses a two-tube amplifier in conjunction with a parallel-tuned resonant tank circuit (see Fig. 4-20). In this circuit, each tube shifts phase 180° so that the total phase shift is 360°, which is equivalent to zero phase shift. We may say that one tube serves as the phase inverting element in place of the RC or LC network which generally performs this function. It is, from an analytical viewpoint, immaterial which tube we choose to look upon as amplifier or phase inverter. The configuration is essentially symmetrical in this respect; both tubes provide amplification and phase inversion. The salient feature of the Franklin oscillator is that the tremendous amplification enables operation with very small coupling to the resonant circuit. Therefore, the frequency is relatively little influenced by changes in tube parameters, and the Q

167

of the resonant circuit is substantially free from degradation. The closest approach to the high-frequency stability inherent in this oscillator is attained by restriction of operation to, or near to, the Class-A region. This should not be accomplished by lowering the amplification of the tubes, but rather by making the capacitors C1 and C2 very small. Additionally, a cathode-follower "buffer" stage is helpful in this regard. Extraction of energy directly from the resonant tank, would, of course, definitely negate the frequency stability otherwise attainable. Obviously, the Franklin oscillator is intended as a low-power frequency-governing stage, not as a power oscillator.

THE BUTLER OSCILLATOR

The Butler oscillator employs a two-tube amplifier with a crystal inserted in the positive-feedback loop embracing the entire amplifier (see Fig. 4-21). In essence, we have a grounded-grid stage driving a cathode follower. Feedback is derived from the cathode of the output tube and is returned to the cathode of the input tube. This feedback connection provides the phase conditions for oscillation without need for phase-shift networks. Accordingly, the crystal oscillates at the frequency at which feedback is maximum. This occurs at its series-resonant frequency because its impedance is then minimum. Indeed, the principal use of the Butler oscillator is found in timing equipment where attaining oscillation at a precise frequency is desired. When the crystal is employed in other oscillator circuits, it generally operates at neither the exact series- nor parallel-resonant frequency. The Butler oscillator is a very con-

Fig. 4-21. Basic Butler oscillator.

venient circuit to construct inasmuch as a duo-triode or triode-pentode within one envelope is admirably suited to the configuration. When it is desired to operate the crystal off-resonance, a vernier tuning range may be obtained by connecting a variable capacitor in series with the crystal. Output may also be derived from the plate circuit of tube V2 by inserting a resistor or choke between the dc supply and the plate. In any event, optimum results will be obtained when a buffer or isolating amplifier is employed.

JUNCTION-TRANSISTOR OSCILLATORS

Practically any electron-tube oscillator can be simulated by a transistorized version of the basic configuration. (A possible exception is the electron-coupled oscillator.) Once the dc biasing arrangement has been given appropriate attention, the transistor is a near substitute for the three element tube. For Class-A amplifiers the relatively low input impedance of the transistor contrasts violently with the extremely high input impedance of the electron tube. However, for Class-C oscillators, both transistor and electron tube display a low input impedance. Actually, the transistor is more suggestive of the pentode tube, notwithstanding the discrepancy in the number of elements. This is so because the collector circuit behaves as a constant-current generator, as does the plate circuit of the pentode. Both output circuits tend to have higher impedances than the plate circuit of the triode.

We see in Fig. 4-22 typical transistor versions of popular electron-tube oscillators. In the ultra-audion type Colpitts, the internal emitter to base capacitance constitutes the second capacitor of the Colpitts capacitive voltage divider. In the tuned-plate/tuned-grid oscillator, the transistor exhibits a "Miller" effect; that is, its input acts as a negative resistance when the output tank is tuned to appear inductive at the parallel-resonant frequency of the emitter-base tank. Although transit-time effects and interelectrode capacitance ultimately impose high-frequency performance limits in electron tubes, a wide frequency spectrum is permissible before attention must be given to these restricting factors. With transistors, however, even at moderate audio frequencies, it is essential to ascertain that the current amplification of the particular transistor type is compatible with the desired oscillation frequency. Germanium power transistors designed for several tens of watts power dissipation rapidly lose current amplification beyond several kilohertz. In contrast, silicon transistors of the diffused base or

A Shunt-fed Hartley; two-battery bias.

B Series-fed tuned-plate/tuned-grid; single-battery bias.

C Colpitts oscillator.

D Ultra-audion type Colpitts.

E Butler oscillator.

Fig. 4-22. Representative transistor oscillators.

"mesa" type are suitable for oscillators generating from several tens to several hundreds of megahertz.

THE UNIJUNCTION TRANSISTOR OSCILLATOR

The unijunction transistor is a switching device in which conduction is abruptly established across two of its three terminals when the voltage impressed across those terminals attains a certain value. Once so initiated, conduction persists until this voltage is reduced to a small value. This behavior is suggestive of a gaseous diode with a very large voltage hysteresis, that is, one in which ionization and deionization voltages differ considerably. Thus, a simple relaxation oscillator employing a unijunction transistor generates a high-amplitude sawtooth voltage wave across the charging capacitor. In the circuit of Fig. 4-23, capacitor C is charged through resistance R1 from the power supply. A small current flows between B1 and B2 and serves the function of back biasing the pn diode that effectively exists between E and B1. As a consequence, negligible conduction prevails between E and B1. However, when the voltage across C attains a sufficiently high value, the distribution of potential gradients within the device is suddenly changed. Specifically, the E-B1 diode section becomes *forward* biased and therefore highly conductive. The capacitor is then rapidly discharged to a voltage low enough to return the E-B1 diode to a reverse-biased condition. It then appears as an open switch and the cycle is repeated as the capacitor again charges toward firing potential.

It is seen that in addition to sawtooth pulses, short-duration spikes are also provided by this simple circuit. These spikes, accessible at B1 and B2, are on the order of several tens of microseconds wide.

So-called "multivibrator" circuits are shown in Fig. 4-24. These circuits provide waveforms similar to those obtainable from two-element oscillators of the Eccles-Jordan type. The essential difference between the circuits of Fig. 4-24 and the relaxation oscillator of Fig. 4-23 is the inclusion of a diode to prevent discharge of the

Fig. 4-23. Unijunction transistor relaxation oscillator.

A Circuit with diode for producing sawtooth waveform.

B Variation of circuit (A).

Fig. 4-24. Unijunction transistor multivibrator circuits.

capacitor through the emitter-Base-1 circuit of the unijunction transistor. Rather, the capacitor must discharge through resistance R4. This requires appreciable time; during this time, the voltage at the emitter relative to Base 1 is sufficient to maintain conduction. When the capacitor is nearly discharged, the voltage at the emitter falls the tiny additional amount necessary to extinguish conduction. A new charging cycle then commences. We see that the diode performs as a hold-off device, enabling the capacitor voltage to trigger conduction, but preventing current from the charged capacitor through the unijunction transistor.

OPTIMIZING THE PERFORMANCE
OF THE MILLER CRYSTAL OSCILLATORS

The Miller-type crystal oscillator shown in Fig. 4-25 is a fantasy in the sense that all of the "add-ons" are not likely to be encountered in a practical circuit. It serves the purpose, however, of illustrating some techniques available for improving the performance of oscillators that may malfunction in some respect, or which do not operate optimally. As shown, such a configuration would be useful for parallel-mode crystals from 100 kHz to ten or 15 MHz.

The rf choke designated by "A" isolates the crystal from the shunting effect of the base-biasing resistors. This is useful in bringing about reliable starting behavior, especially where the crystal may have marginal activity, the beta of the transistor is low, and the biasing resistors are of relatively low value.

The resistors and ferrite beads called out by "B," "C," "D," and "E," are sometimes needed to suppress high-frequency parasitic oscillation. Ordinarily, only one such item is used and it is most likely to be a ferrite bead in the base lead. A resistor, when used, is often in the ten- to 22-ohm range. Ferrite beads are available with different permeabilities and with different dissipative-loss

Fig. 4-25. Band-aids and improvement techniques for a bipolar Miller crystal oscillator. (A) Rf choke in bias circuit makes for easier starting. (B, C, D, E) Resistors and/or ferrite beads discourage high frequency parasitics. (F) Tapped tank inductor can optimize impedance match to transistor. (G) Collector rf choke should be a different value from base-circuit choke A. (H, I, J) Decoupling network discourages low-frequency parasitics. (K) Faraday shield helps isolate oscillator from load.

characteristics so a little experimentation may prove rewarding in discouraging a stubborn parasitic. Basically, the bead and the section of the conductor passing through it form a tiny rf choke. Parasitic oscillations at vhf and uhf frequencies result from stray inductances and stray capacitances and are, accordingly, much dependent upon the constructional practices employed in the oscillator. Also, some transistors (often those with good high-frequency capability) are prone to this type of malfunction.

The tapped tank-coil shown by "F" is an impedance-matching technique that can dramatically improve oscillator operation, especially when outputs exceeding about one-watt are involved. An optimum tap improves the efficiency of both the resonant tank and the transistor. At higher power-levels the collector circuit of the transistor tends to load down the parallel-tuned tank. At "flea-power" levels, such a tap is generally not necessary. Also, other things being equal the tap is more needed for transistor oscillators that operate at low collector voltage and high collector current than for the converse dc operating format.

Attention is directed to the usually-used rf choke, "G," to point out that its inductance should be different from any other that might be incorporated in the oscillator, such as the choke indicated by "A." (Another example of a second rf choke would be found in Colpitts oscillators where a choke is often used in the emitter circuit.) Identically-valued chokes in two parts of an oscillator circuit often cause mysterious oscillation at a lower than intended frequency.

Another source of low-frequency "parasitic" oscillation in bipolar-transistor oscillators comes about from the manner in which such transistors are operated at high frequencies. Intended oscillation usually occurs quite far down on the current-gain characteristic of the transistor. Although the current-gain may be more than sufficient for the intended oscillation mode, the fact remains that much-greater gain is available at lower frequencies. Therefore, if the transistor "sees" any opportunity to oscillate at a low frequency, it will do so. The two possibilities ordinarily encountered are the previously-alluded to "TPTG" oscillations provoked by rf chokes, and relaxation oscillations made possible by appropriate time-constants in the power supply. A tightly-regulated power supply generally inhibits such relaxation-type low-frequency oscillation. Paradoxically, in some cases, it is necessary to sacrifice dc voltage regulation in a special way in order to discourage this type of malfunction. "H," "I," and "J" form a low-frequency decoupling network that often proves effective. At low power-levels, say below one-watt, "H" may range from several-tens of ohms to several hundred of ohms. "I" is an electrolytic capacitor, usually in the vicinity of several hundred microfarads and is "bypassed" at rf by "J," which can be a .01 μF ceramic capacitor. Finally, in order to reduce low-frequency gain, the emitter-resistor bypass capacitor should be no larger than that which suffices for the intended oscillation frequency.

The Faraday-shield designated by "K" is useful where output is derived via inductive coupling. Essentially a grounded electrostatic partition, this technique isolates the oscillator from the capacitive influences of the load. That is, the capacitance between the tank and output coil is no longer effective in reflecting load variations back to the oscillator. Although shown with a crystal oscillator, this technique is especially rewarding when used with self-excited oscillators or VFOs.

Although not shown, a small capacitor connected from collector to base is sometimes needed in the Miller oscillator circuit. This is particularly true for 100 kHz and relatively low-frequency oscillators. Above one MHz, the internal capacitance of the transistor, itself, usually suffices for the needed feedback.

In JFET and MOSFET oscillators, choke "A" is not needed and a ferrite-bead in the gate lead is common practice for parasitic suppression. Also, low-frequency parasitics is not a problem with these devices; accordingly, the decoupling network, "H," "I," "J," has no relevancy.

OPTIMIZING THE PERFORMANCE OF THE COLPITTS CRYSTAL OSCILLATOR

The bipolar transistor and FET Colpitts crystal oscillator are often quickly thrown together and readily yield satisfactory results for many purposes. More often than not, the capacitor divider consists of two equal-valued capacitors. The arrangement tends to work well for a combination of reasons. First, of course, is the relatively high stability provided by a crystal even though the effects of temperature and loading of the active device are ignored. Moreover, the Colpitts circuit is not only an inordinately simple one, but it possesses the inherent feature that harmonics are attenuated in the capacitor divider. Thus, wave purity is easily attained.

Notwithstanding, there are a number of things that can be done to optimize frequency stability. The more important of these can be tabulated as follows.

- Select a good crystal. The "goodness" of a crystal is not altogether governed by its activity or willingness to oscillate. Its temperature coefficient is very important. If the general nature of the temperature coefficient of frequency change is known, it is possible to implement compensatory measures. A crystal with a relatively flat temperature coefficient over the anticipated ambient or operating range is particularly desirable. All things considered, AT-cut crystals operating in the vicinity of 4 MHz tend to have a useful combination of desirable features.

- The use of ordinary-sized, equal-valued capacitors in the capacitive divider is not the best way to isolate the crystal from the active device. Bipolar transistors, especially, have input impedances that vary greatly with temperature and with dc operating current. It therefore makes more sense to make the base-emitter capacitor of the divider three to perhaps five times larger than its mate. In this fashion, the input impedance of the transistor is considerably masked and prevented from unduly influencing the crystal frequency. In practice, the same objective can be accomplished by making both capacitors large and equal. The general idea is to select sizes that have a reactance of about 50-ohms at the frequency of oscillation. This is largely an empirical matter and has to do with the willingness of the circuit to oscillate with very large capacitors—much depends upon the Q of the crystal, the beta of the transistor, and the circuit impedance from emitter to ground.

- Use a buffer amplifier to isolate the oscillator circuit from the driven load.

- Often the series-resonant (Clapp) version of the oscillator contributes toward stability because the crystal then presents a relatively low impedance across its "tank" terminals. This makes this resonance-mode less susceptible to the active device than is the case when the crystal is operated in its parallel resonant mode.

- Use a regulated source of dc operating voltage. Bipolar transistors, especially, undergo wide impedance variations with changing dc voltage.

- Use good wiring and fabrication techniques and make certain that there is high physical and mechanical integrity. These matters are, unfortunately, often overlooked because it tends to be assumed that the crystal and solid-state device take care of everything. However when one considers the several picofarads of trimming capacitance that often suffices to center the crystal frequency, it is apparent that small changes in circuit strays cannot be neglected. Not only is stray capacitance change likely to cause a

Fig. 4-26. Oscillator using special techniques to optimize frequency stability. A 4-MHz AT-cut crystal is series-resonated in a Clapp oscillator circuit which employs heavy capacitive swamping of the transistor input impedance. Oscillator output is derived from a frequency insensitive point and capacitively coupled to a class-A FET buffer. A voltage-regulated power supply is used. Crystal and circuit-temperature coefficient is empirically brought to a low value via optimum combination of silver mica and N750 crystal load capacitors.

frequency deviation, but it does not take too many nanohenries of inductance to do the same. This argues for short leads rigidly secured. Naturally, the best approach is high-quality printed circuitry.

Figure 4-26 shows practical implementation of these suggestions. It incorporates the above-suggested circuitry approaches. Another useful technique might be a constant-temperature oven for the crystal. This suggestion was kept for the last because the crystal oven is too often held to be the panacea for frequency deviation. Its use can dramatically improve stability, but only after the various other influences on frequency constancy are given due consideration. In this regard, ovens using proportionate control rather than on-off temperature cycling are to be preferred. The output of the circuit shown in Fig. 4-26 can be subjected to subsequent power-boosting or frequency multiplication without adversely affecting frequency stability.

Chapter 5

Focusing on
Solid-State Oscillators

FOR THE MOST PART, THE OSCILLATOR CIRCUITS DISCUSSED in the previous chapters are readily implementable as practical oscillators utilizing one or another active device. This, indeed, was the objective of using the concept of the *universal amplifier* in the basic oscillator circuits. Insight into the nature of practical oscillators can be even further enhanced by scrutiny of a number of oscillator circuits utilizing specific active-devices and designated component-values. Accordingly, this chapter will depict and discuss numerous practical oscillators. These will be configured around the various solid-state devices and ICs that are now available and are represented by the symbol of the universal amplifier.

THE UNIVERSAL AMPLIFIER:
A THREE-TERMINAL ACTIVE DEVICE

It would be well to reiterate a few facts pertaining to oscillators and their active devices. Most of the active devices that will be found associated with the oscillator circuits are three-terminal amplifiers. Of course, there are negative-resistance diodes and there are devices with more than three terminals, such as the dual-gate MOSFET. These stand out as exceptions, however. (Even with these devices, the resonant tank in the oscillator "sees" exactly the same kind of negative resistance as is conferred via external feedback paths. Whether one states that an oscillator operates because of positive feedback or because of negative resistance, the two processes are mathematically similar. Also, it is of interest that devices with more than three terminals can be shown to depend upon three "main" terminals or electrodes for their amplifying and oscillating action.)

In actual practice, it is convenient to deal with the fact that most solid-state devices function as three-terminal amplifiers and a few two-terminal devices (diodes) behave as if they already have internal positive feedback paths. Of interest, too, is the fact that circuits can be devised in which any of the terminals of three-terminal devices function as the "common" terminal. Thus, while it is true that the source of a JFET is the "intended" common terminal, either the gate or the drain can be appropriately connected to serve as the common (or grounded) terminal in the oscillator circuit.

It should also be recalled that the active device in an oscillator must provide power gain. It is sometimes erroneously thought that voltage or current gain will suffice. If this were true, an ordinary transformer would be able to provoke and sustain oscillation in a resonant circuit or its equivalent. This power requisite is predicated on fundamentals and applies to negative resistance devices and switching devices, as well as "linear" amplifiers.

100-KHZ TRANSISTOR BUTLER OSCILLATOR

Although better results are now attainable via higher frequency crystals, the 100 kHz crystal has long been used for precise calibration, timing, and instrumentation purposes. Although most of these crystals were intended for tube oscillators, it is usually possible to obtain equally good if not better stability with transistors. This is because transistors are relatively free from aging effects and contribute almost negligible heat to the crystal. For many practical applications, the transistor version of the 100 kHz oscillator can dispense with the crystal oven often used to stabilize against temperature variation. Although transistors, themselves, are subject to thermal effects, the Butler circuit of Fig. 5-1 is minimally affected by small variations about room temperature.

In the Butler oscillator, the crystal oscillates in its series-resonant mode. Usually, these 100 kHz crystals can be pulled into exact frequency through the use of a small series variable capacitor, as shown. Q2 provides a considerable amount of buffering action so that the load does not have as much influence on the oscillation frequency as is often the case with single-transistor oscillators. Notwithstanding, it is generally a good idea to experiment with the output coupling capacitor, making it as small as possible

820

+ 12 V

0.47 μF

2.2 k

1 k

33 k

Out

0.1 μF

100

0.47 μF

12 k

50 pF

100 kHz

6.8 k

5.6 k

Suitable transistors include, 2N2222, 2N697, 2N708, HEP50

Fig. 5-1. 100 kHz Butler oscillator using npn bipolar transistors. Series-mode resonance of the 100 kHz crystal is tunable for calibration purposes.

for the particular application. Inasmuch as the transistors are dc-coupled in this Butler circuit, the wide tolerances of transistor betas ordinarily encountered may require some experimentation with the dc bias on Q2. This can conveniently be brought about by shunting the 2.2 k collector resistor of Q1 with resistances several to many times this value, or by introducing a resistance path from the base of Q2 to ground. The two procedures will produce opposite effects on the base-emitter bias of Q2. The objective of such experimentation is not merely to get operation right for reliable oscillation, but to minimize harmonic production.

From the above, it would appear that harmonic generation, where desired, is readily attainable by biasing Q2 in its nonlinear region. This, however, is detrimental to frequency stability and is likely to adversely affect start-up of the oscillator. It is much better to use subsequent squaring stages to produce harmonic energy. Conversely, the Butler circuit is in its element when it is desired to obtain very good wave-purity.

180

AN EXAMPLE OF A DUAL-GATE MOSFET OSCILLATOR

Figure 5-2 shows three Pierce oscillators utilizing dual-gate MOSFETs. The application is for beat-frequency oscillators enabling the operator of a communications receiver to receive either the upper or lower sideband of a single-sideband signal, or to copy code (cw) messages by means of a beat note. This triple-oscillator circuit is intended for use in conjunction with a 455 kHz i-f channel. In many instances, a single buffer amplifier will probably facilitate implementation.

Note that the two gates of the MOSFETs are connected together. This is the simplest way to use these devices in oscillator circuits and, for practical purposes, yields results comparable to similar circuits using JFETs. Inasmuch as the MOSFET is an inordinately good high-frequency amplifier, the same circuit configuration can be used for higher frequency i-f channels. However, it would then be expedient to reduce the size of the gate capaci-

Fig. 5-2. Example of dual-gate MOSFETS in triple BFO arrangement. Each oscillator is a Pierce circuit. By connecting the two gates together, the operation resembles that of JFETs.

181

tors and of the rf choke. For example, at ten MHz the gate capacitors can be in the vicinity of 30 pF; similarly, the rf choke can be a 500 μH unit. (And, whether or not a buffer is used, the output coupling capacitor would likely be several tens of pF or less.) Other than these changes, it would be necessary to order ten-MHz crystals with the exact frequencies having the same arithmetic displacement from center i-f as the 455 kHz crystals.

The 12-ohm resistors in the drain leads of the MOSFETs discourages vhf and uhf parasitics. An appropriate ferrite bead often suffices for this purpose and in stubborn cases both the resistance and the bead can be used together. Sometimes these suppression techniques are implemented in the gate circuit. In any event, good high-frequency wiring and fabrication techniques should be used. On the other hand, MOSFET and JFET devices are not likely to produce the low-frequency "parasitics" that often plague oscillators using bipolar transistors. (The gain of bipolar transistors is often much higher at audio frequencies than at the operating frequency of the oscillator. This is not true in most MOSFET and JFET oscillator applications.)

At first thought it might appear more straightforward to employ a single active device in which switched crystals provide the three frequency functions. Although this is sometimes done, the precautions necessary (especially for the higher i-f channels) to isolate "hot" crystal terminals from stray capacitances render the single-device design somewhat less than simple to implement. And whereas three active devices would have been awkward and uneconomical with electron tubes, this is certainly not the case in solid-state systems.

SINGLE TRANSISTOR PARALLEL-T OSCILLATOR

The parallel-T (or bridged-T, or twin-T) oscillator shown in Fig. 5-3 uses a single general-purpose npn transistor, such as the 2N2222 or 2N297. This approach represents a simple way to obtain a good sine-wave without resort to large inductors and capacitors. With the T-parameters shown, oscillation occurs at approximately 1 kHz. If other frequencies are desired, the capacitors in the T network should be changed. These capacitors are inversely proportional to oscillation frequency. Thus, if the circuit is to be made to oscillate at 500 Hz, the series-arm capacitors should be doubled to .006 μF and the shunt-arm capacitor should be doubled to .012 μF. The 120 k series-arm elements of the T-network should not be changed because they, in addition to their timing functions, are also involved

Fig. 5-3. A single-transistor parallel-T oscillator. The natural harmonic-discrimination of the T network results in a very low distortion sinusoidal output. The T values shown yield an oscillation frequency of approximately 1 kHz.

in the biasing of the transistor.

The 20 k variable resistance in the shunt-arm of the T-network is adjusted to make the circuit oscillatory. For the 1 kHz circuit shown, this will occur in the 7-11 kilohm range. Note that this shunt-arm resistance would be 60 kilohms according to the parallel-T design equations generally found in texts and handbooks. That is, the shunt resistance is depicted as being one-half the value of the series resistances. However, such a "classic" T will not produce oscillation in the transistor circuit because its amplitude-null is too deep for the gain of the transistor to overcome. As it is, it will be found that higher-beta transistors will be more likely to oscillate, and that some low-beta specimens of a lot will not perform at all. The purpose of the variable shunt-resistance is to find an adjustment where the amplitude null is not too deep and which will still provide the requisite 180-degree phase reversal. Coincidence of these requirements occur over a narrow adjustment range and it will be found that not very much variation of oscillation frequency is feasible.

From what has been said, it can be correctly inferred that this is not a very vigorous oscillator and will not stand much loading. Therefore, an emitter-follower as a buffer amplifier may be found necessary for some applications. On the other hand, the T network is not readily pulled in frequency once its values are optimized for oscillation. Very high beta transistors may oscillate strongly at the expense of wave purity; this can be remedied by inserting a bias current-limiting resistance in series with the base lead. Such a resistance should be bypassed at audio frequencies—a 0.1 μF capacitor will usually suffice. Such an RC combination will have negligible effect on the oscillation frequency. For very low frequencies, consideration must be given to the electrolytic capacitors that may be used. Polarity should be observed and low-leakage types, such as solid-state tantalums, should be selected.

SEVERAL SPECIAL-INTEREST FEEDBACK CIRCUITS

Although a large number of oscillator circuits were explained in the previous chapters, these were by no means all inclusive. They were selected for discussion largely because of popularity of use, and also because of significant differences in configurations and/or mode of operation. Now that most small-signal oscillators are solid-state rather than electron-tube types, several unique oscillator circuits are often seen that would have been awkward to implement with tubes or did not merit much consideration for other reasons when tubes dominated the scene. Although "new" oscillator circuits usually turn out to be derivatives of Colpitts, Hartleys, or Miller circuits, it often happens that the modification, even if slight, contributes significantly to stability, or to some other performance feature.

Three such oscillators that have appeared in technical literature devoted to amateur radio are the Goral, Seiler, and Vackar circuits. They all claim inordinate frequency stability for VFO's (self-excited oscillators using LC resonating circuits). In all three, the resonant circuit is protected in some manner from being degraded in Q, or from being detuned by the active device. This, of course, suggests the long-popular Clapp version of the Colpitts oscillator in which the series resonant tank presents a low-impedance to the input of the active device. However, the three previously-named oscillators use different techniques to accomplish essentially the same result.

The Goral oscillator of Fig. 5-4 has an emitter-follower stage inserted in the feedback path of an otherwise conventional Colpitts oscillator circuit. The midpoint of the capacitive divider (which is

Fig. 5-4. A typical Goral oscillator. The basic idea is to insert buffer amplifiers in both the feedback and the load path. This isolates the frequency determining portion of the oscillator from disturbances.

185

actually part of the resonant tank) now sees a much lower impedance with respect to ground than would be the case without the emitter follower. The power gain of the JFET-bipolar transistor combination is much greater than that of the JFET "oscillator" alone. There is latitude for considerable experimentation in the ratio of the two capacitors used in the Colpitts section of the circuit. This ratio can be optimized for frequency stability without easily running out of feedback. The 100 pF values shown are intended for a first try and will probably ensure oscillation.

Note that the emitter-follower is directly coupled to the JFET. It may be necessary to experiment with bias-determining resistances to ascertain Class-A operation from the emitter-follower. (Also, the output transistor is intended to operate in its Class-A region.) With commonly-used resonant LC combinations, this circuit can perform exceptionally well in the 1-5 MHz region. Those versed in high-frequency techniques will find little difficulty in extending this range. In general, it will pay dividends in frequency stability to strive for a high Q resonant tank via a high ratio of capacitance to inductance.

The Vackar oscillator circuit incorporates a pi-section tank to attain the needed 180-degree phase-reversal in the feedback loop. However, the inverted feedback signal is not directly fed back to the input of the active device; rather, it is loosely coupled through a very small capacitor. Often, a shunt capacitor is introduced to further reduce the coupling. The basic idea is to isolate the resonant circuit as much as possible from the input of the active device, consistent with obtaining reliable oscillation. An example of a Vackar oscillator is shown in Fig. 5-5. Two Class-A buffer amplifiers are used for load isolation. Note the possible use of the shunt capacitor in dotted lines.

This circuit is particularly advantageous with solid-state devices, and especially with bipolar transistors that have inordinately-low input impedances, and that present a widely-varying reactance to the tuned circuit as a consequence of temperature and voltage changes. Tuning can be accomplished by means of a small variable capacitor in parallel with either shunt-arm (C) of the pi network. Although equal values of C are shown, once the overall circuit is operational, these values may be optimized for best stability. Generally, it will be found that the capacitor closest to the base of the transistor can be several-times larger than the capacitor associated with the collector circuit.

It is interesting to note that if the small decoupling capacitor

Fig. 5-5. Example showing the general configuration of the Vackar oscillator. Q1 and its associated circuitry comprise the actual oscillator. Q2 and Q3 are class-A buffer amplifiers. The basic idea of the oscillator is to decouple the resonant circuit from the input of the transistor.

in the feedback loop were shorted out, the operation would be essentially that of the Pierce oscillator, noted for its vigorous feedback and its propensity for harmonic generation, but *not* for frequency stability without the use of a crystal. The introduction of attenuation in the feedback loop (via the small capacitor in the Vackar) prevents over-excitation and effectively isolates the resonant circuit from the active device. In further pursuit of this goal, the output is derived from the emitter, rather than the collector circuit. In this oscillator, it is important that the rf choke in the collector circuit "looks good" at the operating frequency (presents a high inductive reactance). Resonances from distributed capacitance in the choke windings, especially those in the series-resonant mode, can degrade stability or even inhibit oscillation. Ferrite-core chokes are generally suitable for this application. (Sensitivity to rf choke characteristics is common to all oscillator circuits that use chokes for shunt-feeding the dc operating voltage to the oscillator.) Unlike the Clapp, Miller, and the conventional Colpitts oscillators, the Vackar does not readily lend itself to the substitution of a crystal for the LC tank circuit.

The Seiler oscillator has close topological resemblance to a parallel-tuned Colpitts. A basic difference, however, is the location of the dc blocking capacitor conventionally inserted in the base, drain, or grid circuit. In the ordinary Colpitts, this capacitor is electrically close to the active device. In the Seiler version of the circuit, this capacitor is electrically close to the parallel-tuned tank. Moreover, in the Seiler oscillator this capacitor is made no larger than is necessary to reliably start and sustain oscillation.

The practical consequence of this slightly-modified configuration is that greater isolation of the resonant circuit is attained than in the conventional Colpitts. It can be argued that arbitrarily high tank circuit isolation can be achieved in the conventional Colpitts by appropriate manipulation of the size and ratio of the capacitors in the feedback divider network. However, in the Seiler version of the circuit it will be noted that the parallel tank is isolated not only from the active device, but also from the capacitive divider in the feedback network.

Figure 5-6 shows a practical Seiler oscillator circuit. In order to focus attention on the oscillator itself, a buffer amplifier is not shown. However, as with all oscillator circuits from which optimum stability is desired, it will be found profitable to insert a Class-A buffer amplifier between the oscillator output and the load. In this particular oscillator circuit, a JFET source-follower would provide

This capacitor is the "heart" of the Seiler oscillator. Start with .001 μF to ensure the circuit is OK. Then, experimentally reduce the size of this capacitor until it is not much larger than the value needed for reliable oscillation.

+ 12 to + 20 V

.01 μF

MPF102 or equivalent

D

S

25 μH

.001 μF

100 pF

68 k

Output

50 pF

.001 μF

100 μH

Resonant tank shown for 3.6 MHz band.

Fig. 5-6. Example of a practical Seiler oscillator. The potentiality for excellent frequency stability inherent in this circuit is best approached by using a high C-to-L ratio in the resonant tank and by the use of a buffer amplifier and a regulated dc source.

nearly ideal load isolation. Then, if more output voltage or power is needed, a second stage using a bipolar transistor as a common-emitter amplifier would suit the needs of many applications.

It will be noted that the 1N914 or similar diode usually shown in the gate circuit of JFET oscillators is not present in the Seiler oscillator of Fig. 5-6. This is because of the low degree of feedback that results from deliberately making the tank-isolating capacitor barely large enough to maintain oscillation. Under such an operating condition, the pn input diode of the JFET will not be driven into forward conduction and therefore needs no external diode for protection. Because of crystal-holder capacitance, it does not appear that the isolation technique used in the Seiler oscillator lends itself well for use with crystals.

A HARMONIC OSCILLATOR USING
A FUNDAMENTAL- FREQUENCY CRYSTAL

Oscillator circuits are very much like people—it is often the little differences that count. Whereas the circuit of Fig. 5-7 may superficially appear to be just another version of the various configurations already discussed, there is a significant difference in the way

Fig. 5-7. Example of harmonic oscillator. In the above circuit, all frequency multiplication takes place within the transistor and is governed by the resonance of the output tank circuit. The position of the collector tap should be empirically optimized for highest output of the desired harmonic frequency.

it operates. The circuit is basically that of the Colpitts with the apparently minor variation of a tuned-tank circuit in the output. Note that the inductor tap is not arranged in the appropriate way to make a Hartley oscillator, despite a superficial resemblance thereto. In this scheme, the sole purpose of the tap is to provide a better impedance match to the collector circuit of the transistor. This is more important here because the output frequency is at a harmonic of the crystal frequency.

The important aspect of this oscillator is that harmonic energy is delivered from the circuit although an overtone crystal is not used. Indeed, the crystal, a fundamental-frequency, parallel-resonant type, does not vibrate in overtone modes. Although this method of attaining crystal stabilization of higher frequencies is particularly efficient when tuned to operate as a frequency doubler, higher multiplications can be achieved by resonating the collector tank-circuit at the desired harmonic. A small tuning capacitor connected across the crystal will permit greater frequency-pulling than is readily obtained from an overtone crystal.

If one is starting from scratch in the design of a transmitter or receiver, this approach merits consideration for several reasons. Parallel-resonant crystals for fundamental frequency operation are more available than overtone types. Being AT-cut types, they have

190

relatively good temperature stability. This may be true also of third-overtone crystals, but fifth, seventh, and ninth overtone crystals are often BT cuts and tend to have poor temperature-coefficients. As mentioned, fundamental crystals are more amenable to frequency pulling than overtone types. Finally, fundamental crystals are generally less expensive than overtone crystals.

This, however, is not the end of the story. It is possible to combine the two techniques in order to reach a specified frequency. That is, many of the overtone-crystal oscillators can be used in conjunction with an output tank tuned to a desired harmonic. In such instances, the overtone-crystal oscillates at an overtone—usually the third, but further harmonic multiplication takes place within the active device and is selected by the tuned output tank. This can prove a bit tricky and is best attempted with a scheme such as depicted in Fig. 5-8. Here, the emitter inductor, RFC, encourages the crystal to operate in its third overtone mode. at the same time, it discourages vibration in the fundamental mode.

Usually, the oscillators that act also as frequency multipliers exhibit good buffering action with high immunity to frequency pulling from the driven circuit or load.

Fig. 5-8. Overtone oscillator with provision to discourage fundamental frequency oscillation. The inductance of RFC must be large enough to permit operation at f_o, the overtone mode, but small enough to prevent oscillation at the fundamental frequency of the overtone crystal.

191

A BIPOLAR TRANSISTOR
OVERTONE CRYSTAL OSCILLATOR

An overtone crystal oscillator circuit using an npn transistor is shown in Fig. 5-8. In this type of oscillator, the crystal itself oscillates at an odd harmonic of its fundamental mechanical frequency. This is in contradiction to circuits in which the harmonic multiplication takes place within the active element and in which the crystal oscillates at its fundamental mechanical frequency. Indeed, it is often necessary to prevent fundamental-frequency oscillation of the crystal in overtone oscillator circuits. One way of accomplishing this is through the use of an rf choke in the emitter lead, as is seen in Fig. 5-8. The idea is to have sufficient inductance to enable oscillation to occur at the desired harmonic, but insufficient inductance to support oscillation at the fundamental frequency. This generally requires experimentation and may be facilitated via the use of a slug-tuned inductor. The higher the order of the desired harmonic, the easier it is to determine a suitable inductance for RFC, providing one does not run into trouble with stray resonances.

The reason fundamental frequency oscillation is undesirable in the overtone oscillator is that this lower frequency will generally contaminate the output and will tend to ride through the entire receiver or transmitter system. Moreover, the simultaneous oscillation at the fundamental and desired harmonic is generally at the expense of harmonic output level. For many applications, overtone oscillators do not develop a surplus of power, at best. Accordingly, any technique that will result in more efficient production of the desired harmonic is worthy of consideration. Simultaneous oscillation at the fundamental and harmonic frequencies of the crystal is not always immediately obvious and can be a source of spurious frequencies within the system. In the technical literature, one often finds the Pierce oscillator used with overtone crystals because of the strong feedback developed at high frequencies by this oscillator circuit. This, however, can be a two-edged sword because of the readiness at which the crystal will vibrate in its fundamental frequency when used in this simple circuit.

Overtone crystals yield harmonics that are nearly, but not precisely integral multiplies of the fundamental frequency. Moreover, they are not as readily pulled in frequency as are fundamental-frequency crystals. Their use is dictated by the physical and electrical fragility of fundamental-frequency crystals at frequencies higher than 50 or 60 MHz. Just where the demarcation is between fundamental and overtone crystals depends upon the state of the art, cost, the desire to dispense with frequency multipliers, and other factors. Overtone crystals, in any event, find wide application beyond the ten-MHz region, but may also be used at lower frequencies. Keep in mind that overtone crystals are made in such a way as to enhance vibration at a particular odd-harmonic. (Conventional crystals can, but are not likely to perform well in the overtone mode. They are best used with frequency multiplier stages or in oscillator circuits where frequency multiplication occurs within the active element.)

AN OVERTONE CRYSTAL
OSCILLATOR CIRCUIT USING AN FET

Another scheme for exciting overtone modes in overtone crystals is shown in Fig. 5-9. This is not merely an FET substituted for the bipolar transistor of Fig. 5-8. There is more than meets the eye here. First, the Hartley oscillator circuit is not a trivial substitution for the usual Colpitts, for it happens that the Hartley tends to develop stronger feedback in many practical applications. This is favorable for excitation of the higher overtone modes of the crys-

Fig. 5-9. Overtone oscillator using a junction-field-effect transistor. This is a Hartley circuit and is dependent upon series-mode resonance of the crystal at a desired overtone frequency, tuned by the output tank circuit. Optimum efficiency, as well as suppression of fundamental-frequency oscillation is brought about by resonating the crystal shunt capacitance.

tal. At the same time, the gain of the FET does not progressive
increase at lower frequencies as is usually the case with bipol
transistors operating in the radio frequency region. This makes
less likely that the fundamental mode of the overtone crystal w
be excited.

A novel aspect of this circuit is the resonating inductance co:
nected across the crystal. When this inductance has the same nu
merical reactance as the 7 pF or so of holder and socket capacitanc
peak performance in the desired overtone mode is attained. Th:
further discourages any tendency of fundamental-mode oscillatior
Of course, some experimentation is required but this may be facil
tated via the use of a slug-tuned inductor. This oscillator is partic
ularly advantageous for operation at higher than the third overton
of the crystal. Fifth, seventh and ninth overtone modes can be
generated and can be optimized by experimenting with the Q or
the output tank and its feedback-tap. At the higher overtone-modes
best results will be experienced with light output-circuit loading.
Accordingly, a buffer amplifier, preferably another JFET, will lead
to best performance.

THE USE OF DIODES TO
ELECTRONICALLY SELECT CRYSTALS

It is often necessary to be able to switch-select different crystals in a crystal oscillator. This is easy enough to do when one crystal terminal connects to ground. For then, it is only a matter of
wiring the "hot" side of the crystals to the appropriate poles of
a selector switch that has its rotor grounded. This technique, however, does not readily lend itself to situations where it is desired
to select crystal frequencies from a remote point. Also, there are
many oscillator circuits in which both crystal terminals operate at
relatively high impedances above ground. Electronic switching
neatly resolves the problems attending both of these situations. It
also enables the use of a mundane switch rather than a special low-
capacitance type.

Electronic switching is accomplished via the use of small-signal
diodes. Figure 5-10 is an example of electronic switching of nomi-
nally 48 MHz crystals in a modified Pierce oscillator using overtone crystals. The Pierce is not only a good active oscillator circuit
for exciting the overtone crystals, but it also provides the positive
dc connection to the anodes of the switching diodes. The 2.4 k
resistances provide rf isolation as well as limiting the bias current
through the diodes. Thus, the switch can be some distance from

the oscillator. The impedance of the forward biased diode is sufficiently low so as to impart negligible effect on the stability or activity of the oscillator. It is only necessary that the peak rf voltage be much smaller than the bias voltage.

In the oscillator of Fig. 5-10, it will generally be necessary to retune the output tank circuit in order to optimize the overtone frequency of the selected crystal. In other situations, no adjustment beyond actuation of the crystal selector switch is required. For example, a Pierce oscillator can have diode-selectable fundamental-frequency crystals with no need for tuning. (Pierce oscillators using fundamental-frequency crystals often have an rf choke in their output circuit rather than a tuned tank-circuit.)

Fig. 5-10. Example of diode switching of crystals. In the above circuit, the three overtone crystals are in the vicinity of 48 MHz. Selection of a frequency should be accompanied by resonating the output tank circuit.

ELECTRONIC TUNING WITH A
REVERSE-BIASED SILICON DIODE

The effective capacitance evidenced across the terminals of a pn junction is a function of the impressed dc voltage. Unfortunately, the forward-biased situation is also accompanied by an effective dissipative resistance that precludes the use of the capacitance component as part of a resonant tank circuit. On the other hand, virtually no resistive dissipation is associated with the reverse-biased junction, thereby making it a suitable voltage-dependent variable capacitance for tuning purposes. Awareness of this basic feature of pn diodes readily leads to implementations for electronically-tuned oscillators, frequency/phase modulation applications, and remotely-"ganged" rf amplifiers. A few practical aspects of this tuning technique should be kept in mind.

Figure 5-11 shows a generalized oscillator circuit for use in the 2-30 MHz amateur bands. Often, either a specially manufactured varactor or "varicap," or an ordinary pn junction diode will prove satisfactory. Much depends upon the required capacitance range, the demand made on uniformity of the diodes for production purposes, the shape of the capacitance versus voltage tuning curve, and other factors. At higher frequencies a varicap will manifest a higher Q than a common signal or rectifier diode.

Note that the tuning diode connects to the "hot" junction of the Clapp series-resonant tank circuit in Fig. 5-11. This places a heavy burden on the rf choke, the purpose of which is to isolate the dc control circuitry from the tank circuit. The 1.5 k resistor helps relieve some of this burden, but it remains important that this rf choke presents a high inductive reactance over the entire frequency range through which the oscillator is tuned. The control circuitry to the left of the .05 μF capacitor can be of indefinite physical length, but should be shielded to prevent EMI pickup.

The best use of the scheme results from the use of a precision ten-turn potentiometer and digital readout dial. A high-performance voltage-regulated power source is a must, for otherwise the tuning will not be stable and/or hum or noise modulation will be imparted to the oscillator. As is shown, it also makes sense to operate the oscillator from such a supply, too. Other techniques contributing to basic oscillator stability should also be implemented. For example, the Colpitts capacitive divider should utilize silver-mica, or polystyrene capacitors. And, these capacitors should be large enough to effectively "swamp-out" capacitance variations in the active device. (This is of greater importance with bipolar than with

Fig. 5-11. Example of varicap tuning of an oscillator. This arrangement shows the general features of varicap tuning for the amateur HF bands. Various varicaps are available for optimum results, but ordinary signal diodes often work well.

field-effect transistors.) The inductor, L, should have high Q and excellent mechanical integrity. It may be found necessary, in some instances, to connect a several-thousand-ohm trimpot in the ground lead of the tuning potentiometer to prevent the rf voltage from forward-biasing the tuning diode on negative excursions.

WIEN BRIDGE OSCILLATOR

Excellent sine-wave oscillators can be configured around various op amps. This is because op amps provide much greater gain than discrete devices and, at the same time have significantly-higher input impedances. This enables the use of various RC networks without serious loading problems. These oscillators are suitable for low-frequency work, where they dispense with the bulky inductors and capacitors needed for LC oscillators. General-purpose op amps will perform well up to the 100 kHz vicinity and op-amps or voltage-comparators with higher frequency capability can extend this to several MHz, and beyond.

Figure 5-12 shows a practical implementation of the Wien bridge oscillator. In order to optimize the purity of the sinusoidal output wave, a little experimentation is in order so that the amplitude regulating action of the lamp in the negative feedback loop can be maximized. This can often be accomplished by experimenting with values above and below the 1 kilohm resistance shown between the output and the inverting input terminal. Sometimes, a few hundred ohms may be inserted in the ground lead of the lamp. It should be recognized that the primary purpose of the regulation is to minimize distortion. The amplitude stabilization attained by this technique is, of course, desirable, but is not the main goal in most applications of this oscillator circuit. The lamp automatically keeps the net positive feedback low enough so that the op amp works in its Class-A region, thereby avoiding the nonlinear amplitude excursions present in most oscillators.

Instead of lamp filaments, positive temperature coefficient thermistors can be used. These elements are available with smaller thermal time constants than lamps. This feature is important in ultra-low frequency oscillators where the resistance of a lamp filament will change during a single oscillation cycle. If this occurs, the distortion will be increased rather than reduced.

Tuning of this type of oscillator can be achieved by ganging either the "R" or the "C" elements together. Note that the "R" and "C" elements of the series arm can be transposed in order to accommodate a two-gang tuning capacitor with a common metal

$$f = \frac{1 \times 10^6}{2\pi\,RC}\ \text{Hz}$$

Where R is in ohms
Where C is in μF

1 k

Op amp

2
3
6 Sine-wave output

Lamp

R
R C C

Op amp: General purpose such as LM741, LM747, LM748, etc.
Pinout may differ from circuit
Lamp: ELDEMA 1869 or General Electric 1869
Power Supply: $\pm$10-15 volts, with center-tap to ground
For LM741, pin #4 is +Vcc, pin #7 is $-$Vcc

Fig. 5-12. Wien bridge sine-wave oscillator using op amp. Positive feedback for oscillation is provided by the RC network. Variable negative-feedback for amplitude regulation is provided by the lamp filament.

shaft. The rotor of the tuning capacitor must be insulated from ground, however.

THE OP AMP SQUARE WAVE OSCILLATOR

The basic configuration of a widely used multivibrator type oscillator is shown in Fig. 5-13. With the popular 741 op amp, high quality square waves can be generated up to 10 kHz or so. For higher frequencies, there are many op amps available with high switching speed capability. Even better results may be forthcoming from voltage comparators. Depending upon the internal circuitry of the op amp or voltage comparator, a couple of minor variations from the simple circuit of Fig. 5-13 may be necessary. In some instances, a load resistance of a kilohm or so may be needed between the output of the amplifier and the positive dc supply. Also, in some cases a high resistance, say on the order of one-hundred thousand ohms, should be connected between the amplifier's noninverting input (+) and the positive dc supply. (The addition of these two modifications to the circuit of Fig. 5-13 will not appreciably alter its performance and will thus impart universal application to the circuit.) The frequency can be changed by varying either the tim-

Fig. 5-13. A square-wave oscillator suitable for a wide variety of op amps. A 50% duty cycle multivibrator type square-wave generation results from alternate turn on and turn off of the op amp.

ing capacitor or the upper 100-k resistance in the diagram. (The 100-k resistance connected from the amplifier's output to its noninverting input provides the positive feedback path and generally should not be used for frequency manipulation.)

Although the 50% duty cycle square wave provided by the circuit of Fig. 5-13 is required in many applications, other applications are best carried out with other duty-cycle ratios. For such purposes, the basic oscillator of Fig. 5-13 can be readily modified to operate as an adjustable duty-cycle pulse generator. The manner in which this is accomplished is depicted in the circuit of Fig. 5-14. Here, diodes are used to steer the charge and discharge cycle of the timing capacitor through different resistive paths. The 100 k potentiometer enables the adjustment of these resistive paths, and therefore the duty cycle. An interesting and useful feature of this circuit technique is that the pulse repetition rate remains substantially constant over the whole range of duty-cycle ratio. By means of a tapped switch, timing capacitors can be selected to provide useful repetition rates. Otherwise, the statements pertaining to the simpler oscillator of Fig. 5-13 also apply.

The oscillators of Figs. 5-13 and 5-14 are capable of very low frequencies if special attention is given to the electrolytic capacitors which are then required for C. Low-leakage units are needed. Solid-state types are recommended. Frequencies down to and below one-hertz are thereby feasible.

OSCILLATOR USING AN IC TIMER

Although op amps contain many transistors and other simulated components, they are not always the best active device for implementing oscillators. There are a number of ICs which are, in essence, sub-systems and can be as superior to op amps as the op amp is to discrete devices. The kind of superior performance that acrues from these more complex ICs pertains to precision, stability, and flexibility. For example, oscillators made with the popular 555 timer can be designed for a highly predictable frequency, have high immunity to variations of power supply voltage, have internally implemented temperature stability, and have better load isolation than is readily attainable through the use of buffer amplifiers.

The audio-oscillator shown in Fig. 5-15 is intended for hobbyist projects but is easily applicable to demanding instrumentation purposes. This is accomplished simply by paying heed to the stability and quality of the RC timing components. Many of the practical difficulties with LC resonant tanks are circumvented in this type of oscillator. It develops a square wave output. If sine waves are needed, it is easy enough to insert an LC passive filter (either low-pass or band-pass) or an op amp active filter between the output

Fig. 5-14. Adjustable duty cycle pulse generator. Output waveforms for extreme adjustments of duty-cycle potentiometer show that the frequency remains constant. This remains true even when the frequency is changed by changing the timing-capacitor, C.

Fig. 5-15. Audio oscillator using the NE555 timer IC. (A) Circuit of oscillator suitable for code practice and as a general purpose source of audio frequency test voltage. (B) Simplified block diagram of the internal circuitry of the NE555.

(pin 3) and the load. This approach is a compelling one because the filter characteristics are far less critical than would be the LC parameters of a tank circuit in an "ordinary" sine wave oscillator. The formula for calculating frequency in this oscillator is:

$$F = \frac{1.44 \times 10^6}{(R1 + 2R2)(C)}$$

where F is the oscillation frequency in Hz, R1 is as shown in Fig.

5-15 and is in ohms, R2 is as shown in Fig. 5-15 and is in ohms, and C is as shown in Fig. 5-15 and is in microfarads.

For example, if R1 = 5000 ohms, R2 = 100,000 ohms, and C = .01 microfarad, we have:

$$F = \frac{1.44 \times 10^6}{(5000 + 2 \times 100,000)(.01)}$$

$$= \frac{1.44 \times 10^6}{(205,000)(.01)}$$

$$= 704.2 \text{ Hz}$$

Thus, the lowest adjustable frequency of the circuit of Fig. 5-15 is approximately 700 Hz. Higher frequencies are available by adjusting the potentiometer. This is a good frequency range for a code-practice oscillator. Headphones are used, however, this oscillator will drive a 4- or 8-ohm dynamic speaker if a 20 μF electrolytic capacitor is substituted for the 0.1 output capacitor shown.

A SIMPLE FUNCTION GENERATOR

The function generator has been supplementing sine-wave test oscillators and signal generators and threatens to replace these two servicing and calibration instruments. One reason is that the function generator makes available at least three waveshapes, sinusoidal, triangular, and square. There are other reasons, too. The function generator readily allows various sophistications to be incorporated, and is particularly amenable to ultra-low frequency generation. On the other hand, function generators have been extending their high-frequency capabilities since they were first introduced. The result is that a single test instrument now can cover the frequency range previously requiring two instruments—the audio oscillator and the rf signal generator.

Figure 5-16 shows a simple function generator. A frequency range of several Hz to several tens of kHz can be readily covered using inexpensive op amps. With selected op amps, especially with high-speed voltage-comparator types, several hundred kHz can be reached. Commercial function generators can perform out to several MHz and beyond because of special dedicated ICs. The circuit of Fig. 5-16 is "bare-bones;" it enables the experimenter to construct this novel waveform generator with minimal complexity and cost.

The triangle to sine-wave converter is quite rudimentary, com-

$$f = \frac{1 \times 10^6}{2\,(RC)}\ \text{Hz}$$

When R is in ohms.
When C is in μF.

Triangle-wave output

Square-wave output

C

R

10 k

10 k

10 k

1.5 k

Op amps can be 741 type ICs and can operate from a single 15 volt supply.

High-speed voltage-comparators, such as the LM710 series can be used in order to extend high-frequency operation. In this case, a dual power supply is needed for providing +12 V and −5 V.

100 μF

1 k

5.6 k

10 k

1N914

Sine-wave output

Fig. 5-16. A basic function generator using general-purpose op amps. This type of oscillator provides optional outputs of square, triangular, or sine waves.

prising the 1N914 diodes and associated resistors. A triangle-wave contains 15.5% of harmonic energy. These are odd harmonics—the third, fifth, etc. The two-diode sine-wave converter probably reduces this to the vicinity of four- or five-percent. This is, to be sure, a rather impure sine wave. It is, however, sufficiently good for a large variety of test and measurement purposes.

A more serious drawback is the high output impedance of the triangle to sine-wave converter. The impedances often encountered in solid-state circuitry would tend to load down the amplitude of the sine-wave output. An easy solution is to provide a voltage-follower or other type of buffer amplifier so that the sine-wave output will look like a low-impedance source. This was not included in the basic circuit of the function generator because the intent was to focus on the essentials comprising this interesting "oscillator." (In the interest of preventing the load from disturbing the frequency, it may be advisable to provide similar output amplifiers for the triangular and square-wave functions.)

The most convenient method of covering a wide range of generated frequencies is to use a multi-pole switch to select appropriate values of C. R provides the tuning and could be a precision ten-turn potentiometer.

SQUARE-WAVE OSCILLATOR USING LOGIC CIRCUITS

In the ensuing discussions, many oscillators using logic circuits will be dealt with. Like op amps, logic circuits are usually IC modules. Like op amps, logic circuits generally are comprised of many transistors. They even resemble op amps in that many of them have two inputs. (Some, however, have a multiplicity of inputs, or gates.) Being digital devices, logic circuits are intended to be in either an on of an off state so that an oscillator designed around such devices actually makes use of rapid switching action between these two states. Thus, we are not dealing with "amplifiers" in the usual sense. This may appear somewhat puzzling at first, but it should be recalled that similar behavior has been encountered with the op amp square-wave oscillators already described. The real difference in practical terms is that the op amp has a wide linear region between its on and off states. Conversely, the logic circuit has a very narrow "linear" range between its two conductive states. The fact that both circuitries do have a linear range of amplification makes them qualitatively similar square-wave oscillators. The narrow linear range in logic circuits is compensated for by the extremely high gain available within the range.

Thus, from the standpoint of designing oscillator circuits, the logic device operates as a switching amplifier. This being the case, techniques for producing oscillation will still make use of phase inversion, feedback paths, and the cascading of devices. Although an oversimplification, some insight is provided by considering the logic device to function as an over-driven amplifier in oscillator arrangements. A practical feature of logic circuits is that a number of them can be inexpensively utilized in oscillator configurations instead of just a single device.

A typical oscillator using logic circuits is shown in Fig. 5-17. Inasmuch as two cascaded NAND gates are used, the overall loop gain is very high and the square-wave oscillation is vigorous with steep rise and fall times. When we speak of "loop gain," the reference is made to the feedback gain that exists within the very narrow "linear" response of the gates. When the inputs of NAND gates are tied together, the devices function as inverters. This remains true even if one of the inputs is floating because it then assumes a high digital state. However, if one of the inputs is brought low, say by grounding, inverter action does not take place and the circuit is inhibited from oscillating. Thus, switch SW-1 in Fig. 5-17 can be used to turn the oscillations on and off. This can be a useful feature in certain systems, especially when SW-1 is replaced by an electronic switching technique.

Although an audio frequency application is depicted, operation

Fig. 5-17. NAND element square-wave oscillator for the 1 kHz region. Actually, there is only one pin 14 and only one pin 7 on the SN7400 module, which contains four NAND gates. Only two of these gates are used in the above oscillator.

may be extended to several megahertzs or higher by appropriate selection of the RC timing parameters.

A FEW WORDS ABOUT THE SN7400 NAND GATE IC

The circuit, pin-connections, and truth table shown in Fig. 5-18 provide useful information for the use of the popular SN744 TTL quad NAND gate. Circuits configured about this, and other logic devices, generally omit power supply connections. However, the pin-connection diagrams shows that pin 14 is the + 5 volt VCC connection and that ground connects to pin 7. This technique simplifies schematics, enabling one to confine and focus interest on the essentials of circuits often comprised of numerous interconnections.

From the circuit of this device, one can infer that considerable gain is available, greater than can readily be obtained from a discrete device. Although such logic devices make speedy transitions between their on and off states, the gain during these transitions is available as feedback in oscillator circuits. Even when crystals or LC resonant tanks are used for frequency determination and stabilization, the output of such logic devices as the SN7400 is a square wave. Fortunately, this is the required waveshape for clocking logic systems. For other purposes, the square wave may be converted to a sine-wave with a resonant circuit or filter. One of the features of using logic gates for clock oscillators is that the voltage levels are compatible for the operation of the clocked logic systems.

The truth table reveals that the SN7400 becomes an inverter

if the gates are tied together. Thus, the device will simulate the characteristics of a common-emitter transistor. An unconnected gate tends to assume the high state. Assuming, then, that such an unconnected gate is always high, the truth table indicates that the device will still work as an inverter. The oscillator of Fig. 5-17 should therefore operate in essentially the same manner whether the extra gates are unconnected, or tied to the active gates. In actual practice it may pay to experiment with these connections especially at higher frequencies.

INPUTS		OUTPUT
A	B	
0	0	1
0	1	1
1	0	1
1	1	0

Fig. 5-18. Circuit, pinout, and truth table of the SN7400 NAND gate. This device is typical of logic ICs used in oscillators. (A) Internal circuit of the SN7400. Note dual-emitter input transistor. (B) Pinout of the SN7400. Note common power supply provision. (C) Truth table.

207

LOGIC CIRCUIT SQUARE-WAVE
OSCILLATOR WITH CRYSTAL STABILIZATION

The crystal oscillator shown in Fig. 5-19 embodies the same oscillation principle as the low frequency square-wave oscillator of Fig. 5-17. Thus, if the crystal is shorted out, a self-excited oscillatory mode occurs in which the frequency is governed by the 33 pF trimmer capacitor. In some instances, a higher free-running frequency is generated when the crystal is removed from its socket because of the stray capacitance of the holder. But this should not be inferred to signify instability when the crystal is in place; for then, the oscillation frequency is firmly that of the crystal. Inasmuch as there is no tuned tank circuit, other crystals than the one-megahertz unit indicated may be used. Regardless of the type of crystal used, series-mode resonance is involved in this circuit. Even an overtone crystal will likely oscillate at its series-resonant fundamental mode. Feedback is very strong and crystals of marginal activity tend to be sure-fire in this simple oscillator.

The third NAND gate functions as an output buffer amplifier. It, too, operates in the off-on manner of logic devices and ensures a clean square wave for driving subsequent logic. At the same time, it increases the fan-out that would be available directly from the oscillator circuit itself.

An interesting and useful aspect of oscillators of this type is that they are rich in odd-harmonic energy. Thus, they are immedi-

U1-A, U1-B, and U1-C are three sections of an SN7400 quad NAND gate

Fig. 5-19. Crystal oscillator using NAND gates. The circuit includes a buffer output stage and delivers a square waveform.

$$f = \frac{1 \times 10^6}{2\,(RC)}\ Hz$$

where R is in ohms
where C is in μF

ICs are two sections of quad-NAND gate SN7400
Transistors are 2N2222, or similar "workhorses."
Pin 7 on the SN7400 connects to ground and pin 14 connects to +5 V

Fig. 5-20. IC clock oscillator with complementary outputs. For proper operation, the "R" values should be in the vicinity of 100 ohms. The "C" values have considerable latitude and should be selected to determine frequency.

ately useful as frequency markers. Also, by means of appropriate tuned circuits, a harmonic frequency can be "sucked out" of the output waveform and amplified or processed for a specific application. This is a slightly different technique than that provided by frequency multiplier stages in "linear" rf systems. This circuit should work well up to ten or 15 MHz, which is close to the practical limit for nonovertone AT-cut crystals. Incidentally, even though the circuit generates square waves, the crystal oscillates sinusoidally.

A CLOCK OSCILLATOR FORMED
FROM CROSS-COUPLED TTL NAND GATES

Cross-coupled gates are frequently encountered in logic circuits. This configuration is basically that of the family of multivibrators used for many purposes with discrete devices. Logic gates often have very high gains and, if properly used, tend to be superior to discrete-device multivibrators. Thus, the arrangement shown in Fig.

5-20 is an astable multivibrator with transistor output amplifiers connected to provide complementary outputs. This is an excellent clock oscillator for general use with logic systems.

The SN7400 IC being a quad NAND gate utilizing TTL logic levels, should be operated from a five-volt regulated power supply. The oscillation frequency is given by

$$\frac{1}{1 \ (RC)}$$

In order that the square-wave generated should have a symmetrical duty-cycle, it is assumed that the "R" values are equal; likewise, the "C" values are equal. Also, the "R" values should be in the vicinity of 100 ohms in order to advantageously use the threshold characteristics of the gates. This implies that frequency determination is best done via manipulation of the "C" values. With the R values equal to 100 ohms, and the C values equal to 0.1 μF, the circuit of Fig. 5-20 provides a 50-kHz square wave from each of its output terminals with respect to ground. The complementary phase relationship between the two clock frequencies makes this arrangement particularly useful for clocking logic circuits.

Although the unused gates are left disconnected in this circuit, it is common practice to tie the gates together; that is, pin 1 would connect to pin 2, and pin 13 would connect to pin 13. (Although this is optional here, due consideration must be given to unused gates in many logic circuits.)

MORE EXAMPLES OF LOGIC DEVICE OSCILLATORS

Four additional logic IC oscillators are shown in Figs. 5-21, 5-22, 5-23, and 5-24. These, together with those already depicted, amply demonstrate the diversity of such oscillator configurations. A common denominator of these systems is that high levels of feedback are available. Start-up troubles that sometimes plague discrete-device oscillators tend to be absent in these arrangements. Also, they are basically square-wave oscillators because of the off-on operational mode of logic ICs. Of course, good sine-wave output may be obtained via an appropriate low-pass or band-pass filter. This is not often encountered because these oscillators are most often used to drive other logic devices that function best on square-wave inputs.

Probably the simplest of these oscillator systems is the NOR-gate crystal-oscillator of Fig. 5-21. Although the output NOR gate

Fig. 5-21. NOR gate crystal oscillator for supplying 3.579 MHz color TV frequency. These crystals restore the sub-carrier frequency in TV sets.

is not needed to square-up the signal, it provides buffer action and delivers a cleaner wavetrain than is available directly at the output of the first IC, i.e., the oscillator proper. It also provides increased fan-out capability by preventing needless loading of the oscillating portion of the circuit.

Another crystal oscillator arrangement is shown in Fig. 5-22. This one utilizes five devices in a hex-inverter IC. This is an ex-

Fig. 5-22. 100 kHz oscillator using five sections of a CMOS hex inverter. Circuits of this kind employing several cascaded stages develop high feedback and can sometimes generate self-excited oscillations if the crystal is defective or is not in place.

Fig. 5-23. Crystal oscillator using Schmitt trigger logic elements. This is a vigorous square-wave oscillator. The function of C_x is to discourage generation of spurious frequencies.

ceptionally strong oscillator and is well-suited for frequency-marker generators. The frequency-adjust provision is somewhat unconventional. The trimmer capacitor instead of being inserted directly in series with the crystal operates in an overall negative-feedback loop. This tends to make the positive-feedback loop containing the crystal less sensitive to stray capacitance, thereby increasing the pulling range of the oscillating frequency. Spurious or noncrystal

Fig. 5-24. 1 kHz complementary output oscillator using MOS NOR gates. The square-wave outputs are 180 degrees out of phase.

governed frequencies can be inhibited by experimentation with the resistor (10 k) connected between the first and second stages. It should be noted that the last stage is not truly a buffer and it may be found profitable to utilize the sixth inverter in this IC as an output stage for better load-isolating action.

The arrangement shown in Fig. 5-23 is an interesting one employing Schmitt-trigger elements of a logic IC. Here, too, spurious frequencies may be excited by the vigorous feedback. A value of capacitor C_x may be empirically determined which will attenuate or inhibit spurious frequencies, but which will still allow reliable start-up of the intended crystal frequency. As with the previous oscillator circuit, it may be found desirable to add an additional stage to perform as a buffer amplifier.

The MOS NOR-gate oscillator depicted in Fig. 5-24 is especially useful for logic systems requiring complementary clocking. It will be seen that the output waveforms of this oscillator are displaced in phase by 180 degrees. Insight into the nature of this oscillator scheme may be clarified by noting that the NOR gates behave as inverters because their gates are connected together.

VOLTAGE-CONTROLLED OSCILLATORS

Voltage-controlled oscillators have many important uses. A few of these are found in phase-locked loops, instrumentation, such as the voltage-to-frequency converters in some digital multimeters, tuning in communications equipment, certain switching regulators, telemetering, and in remote control techniques. Two general approaches are common—one for radio frequencies and the other for audio frequencies. A voltage-dependent capacitor is used for the radio frequency range as part of a resonant circuit. Such a voltage-dependent capacitor is usually a reverse-biased junction diode, or varactor. Because negligible bias current flows, such a simulated capacitor exhibits low loss and the overall Q of the resonant circuit can be very high, being governed mostly by the inductor. This type of voltage-controlled oscillator has already been discussed. A typical circuit is that of Fig. 5-11.

For audio frequencies, varactor techniques can also be used, but there are certain practical difficulties, one being the outsized physical size of the tuned inductor. Another is the relatively low capacitance of pn junction diodes. Of course, there are low-frequency oscillators that do not use inductors, and it is possible to parallel a number of varactors. But most of the audio-frequency voltage-controlled oscillators encountered in practice make use of some type of relaxation oscillator or multivibrator in which either resistance or voltage level is manipulated by the control voltage. Figure 5-25 shows a unijunction oscillator in which the charging rate of the timing capacitor, C, is altered by a dc control voltage. Here, the bipolar transistor is connected to act as a voltage-dependent resistance shunting the timing capacitor. The control range extends from about +0.8 to 2.5 volts with the frequency decreasing with increasing control voltage. The component values are suitable for the mid-audio frequencies and the actual range of controlled frequencies is determined by selection of capacitor C.

An alternate way of voltage-tuning a unijunction oscillator is shown in Fig. 5-26. In this circuit, the bipolar transistor is in series with the timing capacitor and thereby takes the place of the 50-k charging resistance used in Fig. 5-25. In this arrangement, the frequency increases with increasing dc control voltage. The biasing resistance, R_x can be empirically determined to produce an optimum relationship between applied voltage and resulting frequency. Its value will depend on the beta of the 2N2222 transistor and the nature of the variable dc voltage source. The 2 μF output coupling capacitor enables the oscillator to drive a small 4- or 8-ohm dynamic speaker.

Another type of VCO (voltage-controlled oscillator) is shown in Fig. 5-27. This is essentially a multivibrator circuit much used for square wave generation. Here, however, the basic circuit is modified to allow control of the voltage level that the single feed-

Fig. 5-25. Simple voltage-controlled oscillator using a unijunction transistor. The bipolar transistor is used as a variable resistance to delay the charging rate of timing-capacitor C.

Fig. 5-26. Another unijunction VCO circuit. In this arrangement, the bipolar transistor acts as a voltage-controlled resistance in the series-charging path of the timing capacitor. Experiment with R_x for best results; 330 k is a good starting value.

back capacitor "sees" during its charge cycle. The higher the positive control voltage, the less time must be consumed in the charging cycle and the higher is the output frequency of the oscillator. Some experimentation may be in order with regard to the two resistors in order to optimize the control range. Within limits, decreasing the value of the input resistor (shown as 220 k) will improve the control range, particularly on the high-frequency side. It is always best to operate circuits of this type from a regulated dc source so that only the dc control voltage can change the frequency. Although MOS inverters are shown, the basic idea can be extended to other devices such as op amps and comparators. Also, it may prove feasible to use this scheme well into the moderate radio frequencies, say in the several-MHz range.

Fig. 5-27. Audio frequency voltage-controlled oscillator using MOS inverters. This circuit technique leaves four inverters for other circuit or systems functions.

215

Fig. 5-28. A VCO derived from a CMOS phase-locked loop IC. This dedicated IC provides pin connections so that the internal VCO may be used independently.

For applications where the output frequency of an oscillator must be a very linear and precise function of the dc control voltage, it is not wise to attempt to achieve this with discrete devices or even with op amps. Semiconductor firms now make dedicated ICs especially designed for such high performance. These may be called either voltage-controlled oscillators or voltage-to-frequency converters. Two such ICs that yield exceptional linearity are the National Semiconductor LM131 and the LM566. Also, Siliconix makes the LD111, a voltage-to-frequency converter intended for digital multimeters.

Another way to obtain better performance from a voltage-controlled oscillator than is readily feasible by "rolling your own" is to make use of the VCO contained on the chip of a phase-locked loop IC. Fortunately, in certain ICs of this type, the VCO section of the loop may be used independently. A dedicated IC of this type is the 4046 CMOS phase-locked loop. Figure 5-28 shows how a simple VCO can be configured by appropriate selection of the IC pins. In this arrangement, the frequency range is governed by the RC time-constant of the resistor connected between pins 11 and 8, and by the capacitor connected between pins 6 and 7. With the indicated component values, the frequency is controlled from about 1 Hz to 10 kilohertz. This may be extended to about 1 megahertz by appropriate selection of the RC time constant. However, the resistive component of the RC value should be greater than 10 kilohms for reliable oscillation. Despite the tremendous control range provided by this sub-system, the integrity of the 50% duty cycle square wave is maintained.

Index

Edited by Roland S. Phelps